Sustainable Advanced Manufacturing and Materials Processing

This book encapsulates and highlights the most recent innovations, breakthroughs, and comparisons of advanced sustainable manufacturing and material processing techniques for high-performance materials applications with a focus on sustainability and using conventional available methods.

Sustainable Advanced Manufacturing and Materials Processing: Methods and Technologies addresses the various sustainable manufacturing and materials processing techniques for advanced materials. It discusses advancements in conventional and non-conventional techniques used in casting, joining, drilling, surface engineering, sintering, and composite manufacturing. The book focuses on a wide range of manufacturing techniques and materials processing technologies along with their benefits, limitations, and sustainability quotient. The conventional and advanced processes are compared in parallel to understand the need for advanced methods in manufacturing technology.

This book is helpful to academic scholars and commercial manufacturers in giving them a first-hand source of information on sustainable manufacturing and material processing technology.

Sustainable Manufacturing Technologies: Additive, Subtractive, and Hybrid

Series Editors: Chander Prakash, Sunpreet Singh, Seeram Ramakrishna, and Linda Yongling Wu

This book series offers the reader comprehensive insights of recent research breakthroughs in additive, subtractive, and hybrid technologies while emphasizing their sustainability aspects. Sustainability has become an integral part of all manufacturing enterprises to provide various techno-social pathways toward developing environmental friendly manufacturing practices. It has also been found that numerous manufacturing firms are still reluctant to upgrade their conventional practices to sophisticated sustainable approaches. Therefore this new book series is aimed to provide a globalized platform to share innovative manufacturing mythologies and technologies. The books will encourage the eminent issues of the conventional and non-conventional manufacturing technologies and cover recent innovations.

Advances in Manufacturing Technology
Computational Materials Processing and Characterization
Edited by Rupinder Singh, Sukhdeep Singh Dhami, and B. S. Pabla

Additive Manufacturing for Plastic Recycling
Efforts in Boosting A Circular Economy
Edited by Rupinder Singh and Ranvijay Kumar

Additive Manufacturing Processes in Biomedical Engineering
Advanced Fabrication Methods and Rapid Tooling Techniques
Edited by Atul Babbar, Ankit Sharma, Vivek Jain, Dheeraj Gupta

Additive Manufacturing of Polymers for Tissue Engineering
Fundamentals, Applications, and Future Advancements
Edited by Atul Babbar, Ranvijay Kumar, Vikas Dhawan, Nishant Ranjan, Ankit Sharma

Sustainable Advanced Manufacturing and Materials Processing
Methods and Technologies
Edited by Sarbjeet Kaushal, Ishbir Singh, Satnam Singh, and Ankit Gupta

For more information on this series, please visit: www.routledge.com/Sustainable-Manufacturing-Technologies-Additive-Subtractive-and-Hybrid/book-series/CRCSMTASH

Sustainable Advanced Manufacturing and Materials Processing

Methods and Technologies

Edited by
Sarbjeet Kaushal, Ishbir Singh,
Satnam Singh, and Ankit Gupta

CRC Press
Taylor & Francis Group
Boca Raton London New York

CRC Press is an imprint of the
Taylor & Francis Group, an **informa** business

First edition published 2023
by CRC Press
6000 Broken Sound Parkway NW, Suite 300, Boca Raton, FL 33487–2742

and by CRC Press
4 Park Square, Milton Park, Abingdon, Oxon, OX14 4RN

CRC Press is an imprint of Taylor & Francis Group, LLC

© 2023 Taylor & Francis Group, LLC

ISBN: 978-1-032-21626-3 (hbk)
ISBN: 978-1-032-21627-0 (pbk)
ISBN: 978-1-003-26929-8 (ebk)

DOI: 10.1201/9781003269298

Typeset in Times
by Apex CoVantage, LLC

Contents

Editor Biographies

Dr. Sarbjeet Kaushal is currently working as an associate professor in the Department of Mechanical Engineering at Gulzar Group of Institutes, Khanna, Ludhiana, Punjab, India. He received his Ph.D. in mechanical engineering from the Thapar Institute of Engineering and Technology, Patiala, India, in 2019. He has published more than 30 papers in refereed journals, including 20 SCI/SCIE indexed journals, and is also a reviewer of several peer-reviewed journals. Dr. Kaushal's research interests include surface engineering, microwave processing of materials, tribology of composite materials, and advanced manufacturing processes.

Dr. Ishbir Singh is currently working as the Director of Engineering, Gulzar Group of Institutes, Khanna, Ludhiana (Pb), India. He received his Ph.D. in mechanical engineering from the Thapar Institute of Engineering and Technology, India. He has more than 18 years of experience in industry, academia, and research and has served in many prestigious positions like head of department, chief warden, associate dean, project in-charge, faculty advisor, and so on. Dr. Singh has received several appreciation letters and commendation certificates for his contribution in the field of technical education and is an active researcher in the fields of design, biodynamics, CAD/CAM/CAE, finite-element method (FEM), and human body vibrations. He has several publications to his credit in national and international conferences and journals of repute. He is an author and reviewer for high-impact and reputed journals and has filed several patents.

Dr. Satnam Singh has worked as an assistant professor in the Department of Mechanical Engineering at the NorthCap University, India, since 2016. He received his Ph.D. in mechanical engineering from Thapar Institute of Engineering and Technology, Patiala, India, in 2017. He has more than nine years of experience in teaching and industry. He has published more than 25 SCI/SCIE/SCOPUS papers in various reputed international journals and has presented papers at international conferences. He has applied for three Indian patents and is actively working in the field of microwave materials processing technologies. His fields of research include microwave materials processing, development of composites (MMCs, PMCs and CMCs), surface engineering, and joining of materials.

Dr. Ankit Gupta received his master's degree in CAD/CAM and robotics from the Indian Institute of Technology Roorkee, India, and a doctoral degree from the Indian Institute of Technology Mandi, India. He has published more than 50 papers in refereed journals, including SCI/SCIE/SCOPUS-indexed journals and conference proceedings. He is a member of the editorial board of several peer-reviewed journals. Dr. Gupta's research interests involve uncertainty quantification, deformation theory, structural response, imperfection sensitivity, and nanotechnology.

Contributors

Gaurav Arora
Chandigarh University
Mohali, Punjab, India

Samson Mekbib Atnaw
Addis Ababa Science and Technology
   University
Ethiopia

Ravi B
Swarna Bharathi Institute of Science &
   Technology (SBIT)
Khammam, Telangana, India

Anmol Bhatia
Delhi Technological University
Delhi, India
and
The NorthCap University
Gurugram, India

Melaku Desta
Addis Ababa Science and Technology
   University
Ethiopia

Ankit Gupta
Shiv Nadar University
Greater Noida, Uttar Pradesh, India

Suresh K
Annamalai University
Chidambaram, Tamil Nadu, India

Sarbjeet Kaushal
Gulzar Group of Institutions
Ludhiana, India

Jaswinder Kumar
Punjab State Aeronautical Engineering
   College
Patiala, India

Subodh Kumar
Madanapalle Institute of Technology &
   Science
Madanapalle, India

Sarina Lim
University of Putra
Malaysia

Elias G. Michael
Addis Ababa Science and Technology
   University
Ethiopia

Sivaprakasam Palani
Addis Ababa Science and Technology
   University
Ethiopia

Ravi Shankar Prasad
JK Lakshmipat University Jaipur
Jaipur, India

Saurabh Rai
Shiv Nadar University
Greater Noida, Uttar Pradesh, India

Sachin Saini
RIMT University
Mandi, Gobindgarh, India

Saloni
Gulzar Group of Institutions
Ludhiana, India

Doordarshi Singh
BBSBEC
Fatehgarh Sahib, India

Ishbir Singh
Gulzar Group of Institutions
Ludhiana, India

Jashanpreet Singh
Punjab State Aeronautical Engineering
    College
Patiala, India

Kulwant Singh
Chandigarh University
Mohali, Punjab, India

Rajwinder Singh
Thapar Institute of Engineering and
    Technology
Patiala, India

Satnam Singh
The NorthCap University
Gurugram, Haryana, India

Harsimran Singh Sodhi
Chandigarh University
India

Mohammod Toseef
Punjab State Aeronautical Engineering
    College
Patiala, India

Reeta Wattal
Delhi Technological University
Delhi, India

1 Introduction to Sustainability of Manufacturing and Materials Processing Technologies

Satnam Singh

CONTENTS

1.1 INTRODUCTION

Manufacturing sectors globally are being transformed and upgraded due to strict norms from various government policies. Energy savings and emission reduction (ESER) is a globally adopted strategy by industries to promote sustainability goals (Cai et al., 2019; Herrmann et al., 2014). The global strategy for any manufacturing industry to grow and sustain is to focus on energy-efficient processes, green manufacturing, and low carbon emissions. All sectors are being reformed due to directives from government policies for lower carbon footprints and energy savings. Out of all the sectors, manufacturing plays a crucial role in the economic development of any nation. This sector is now on the radar for transformation from conventional methods to advanced technologies, where sustainability is one of the major parameters. Sustainability should further provide room for clean, green, and low carbon–emission processes. Now the major question arises of how industries are going to innovate and implement various strategies for green transition while addressing sustainability issues. Sustainability issues in the manufacturing area have been a major topic of discussion at the global level, where researchers are considering sustainable manufacturing an effective solution for continuous growth and development (Böhringer & Jochem, 2007). Over the years, lean manufacturing or lean production methods have provided a route for sustainable and cost-effective manufacturing policies. A lot of literature is available on the application of lean practices for sustainable development in the core area of manufacturing where a lot of methods, modeling, analysis, and

DOI: 10.1201/9781003269298-1

frameworks have been reported (Martínez León & Calvo-Amodio, 2017; Abualfaraa et al., 2020). Lean practices have been used in almost all sectors, including automobiles, foundries, hospitality, supply chains, education, and so on (Hallam & Contreras, 2018; Comm & Mathaisel, 2005; Arumugam et al., 2022).

However, sustainability is still a very broad term that needs to be analyzed at the root level for any manufacturing process. Further, there can be multiple parameters that can contribute to the sustainability of any manufacturing/material processing technology. These parameters need to be selected carefully to understand their impact on the overall sustainability of processes, and mathematical models can be developed to understand this in a better way. Lucato et al. (2018) reported a mathematical model for understanding the sustainability of a computer numerical control lathe machining process considering three major dimensions: environmental issues, economic performance, and social gains. Such a study can be associated with every process to understand the sustainability index of manufacturing processes. There are still limitations to applying parametric analysis on all the available manufacturing and materials processing technologies and find the sustainability index. Many ideas and initiatives have been developed to study sustainability over the years, but some limitations still exist, such as:

- The sustainability framework is different for different sectors and cannot be universally applied.
- Lean practices/frameworks work for the industry as whole but cannot be segregated for the manufacturing process alone.
- Measurement for sustainability is carried out on various parameters and may consider environmental, economic, and social parameters separate variables, or they may be integrated with one another.
- The concept may be too complicated to apply for the industry to evaluate all its processes.
- Developed frameworks may provide data that may be difficult to implement to achieve the objectives of sustainable development.

Hence this chapter proposes a simple mathematical model to evaluate the manufacturing process based on simple measurable parameters to identify the sustainability index. The parameters selected for this model are general and can be identified easily and used per requirements.

1.2 PROPOSED MODEL FOR SUSTAINABILITY INDEX

Manufacturing processes have various measurable parameters that can easily be determined by the industries. Some of these parameters are shown in Figure 1.1.

The parameters are common in nature and can be associated with any manufacturing/ materials processing technology. These parameters are based on four common points:

- Parameters determining the economic performance of processes
- Parameters affecting the environment
- Parameters affecting process capabilities
- Safety parameters

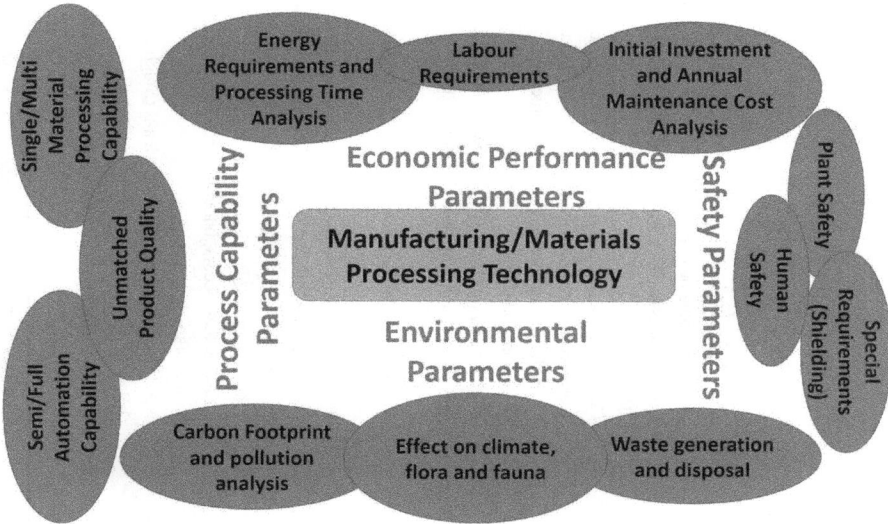

FIGURE 1.1 Common measurable parameters associated with any manufacturing process.

These four common parameters can further be divided into subparameters; for example, the economic performance of any process can be directly measured by the energy consumption in processing/manufacturing. This is directly related to time consumed to manufacture a product. Further, the economy of a process can also be related to the initial investment made and annual maintenance cost for the process. The life cycle of the equipment used in the process is also an important parameter. On similar lines, environmental safety can be linked to carbon emissions, pollution, and effects on flora and fauna. Capability parameters may be linked to the capability of a process to handle multi-material systems, the extent to which the process can be automated, and whether the quality of products produced by the process is unmatched and cannot be obtained by other process. Safety parameters may be linked to special requirements (shielding due to explosive nature or radiation, etc.), human safety, and plant safety. All these parameters can be ranked from 1 to 5 based on available benchmarking data or a user-designed matrix. Based on the selected parameters and subparameters, Equation 1 can be used to design the sustainability index value.

$$SI = f\left(P_1, P_2, P_3 \ldots\ldots\ldots\ldots P_n\right) \tag{1}$$

where SI is the sustainability index value, P is a parameter associated with the selected manufacturing process, and f is a function of parameter scores that can be based on the relative scores obtained by comparison with conventional system (see Tables 1.1 and 1.2). To understand the process, let us consider a case study of a microwave materials processing process and some parameters.

Let the SI value be between 1 and 5, where 1 is the lowest sustainability index value, which indicates the process is not sustainable and will not be able to survive,

TABLE 1.1

Parameters Selected for Sustainability Index for Microwave Sintering Process

S. No.	Parameter Category	Parameter Selected	Range	Relevance in Comparison to Conventional Sintering Process	Score	Justification
1	Economic considerations	Energy required	1–5	The microwave sintering process is a well-established route for ceramics with very high energy savings and lower processing times.	5	More than 90% savings in energy in comparison to conventional heating system (Singh et al., 2015)
2		Processing time			5	Lower processing time by factor of 10 (Singh et al., 2015)
3		Initial cost and annual maintenance cost			4	Cost is comparable to conventional furnaces but highly cost effective in comparison to advanced technologies such as laser sintering processes (Pathania et al, 2015)
4	Process capabilities	Product quality	1–5	Microwave heating does not promote high grain growth due to lower processing time and also suppresses defects.	4	High-quality products with the fewest defects are reported in the literature (Singh et al., 2015)
5		Material system		Microwave heating can be used for ceramics, metallic powders, plastics, glass, and so on.	4	Versatile and can be used for all types except bulk metals
6		Automation		Difficult to automate, as research is ongoing in this area. Lower volume in comparison to conventional furnaces.	3	No work focusing on fully automatic materials processing systems has been carried out, but it can be achieved
7	Safety	Human safety	1–5	Highly safe process and threat until direct exposure of tissues.	4	There are only safety issues if a microwave radiation leak occurs, but commercial systems are well designed
8		Plant safety		No threat to plant, as microwave systems are safe with proper shielding, in comparison to conventional processes, which need shielding from radiation leaks.	3	Need shielding from microwave leaks (Singh et al., 2015)

TABLE 1.2

Parameter wise Calculation of Scores for Microwave Sintering Process

S. No.	Parameter Category	Parameter Selected	Individual Score	Average Category Score	Overall Process Score	Sustainability Index Value
1	Economic considerations	Energy required	5	4.67	11.84	3.95
2		Processing time	5			
3		Initial cost and annual maintenance cost	4			
4	Process capabilities	Product quality	4	3.67		
5		Material system	4			
6		Automation	3			
7	Safety	Human safety	4	3.5		
8		Plant safety	3			

whereas a value of 5 is the highest value and represents the process being highly sustainable per the selected parameters. Table 1.1 shows the various parameters, their relevance, and the ranges for the calculation of the sustainability index for the microwave sintering process of ceramics.

The scores from the range of values are selected based on the user experience and comparisons between the selected parameters of conventional and microwave sintering processes. The benefits of microwave heating in terms of energy savings and lower processing time are known globally and hence can get a full score. Similarly, other parameters are scored on the basis of available literature and user experience. The calculation of final scores for the parameter category is shown here:

$$\textbf{\textit{Average Score}}_{(Economic\ Consideration)} = (Score\ \Sigma_{i=1}^{n} P_i)/n$$
$$= (5 + 5 + 4)/3$$
$$= 14/3 = 4.67$$

$$\textbf{\textit{Average Score}}_{(Process\ Capabilities)} = (Score\ \Sigma_{i=1}^{n} P_i)/n$$
$$= (4 + 4 + 3)/3$$
$$= 11/3 = 3.67$$

$$\textbf{\textit{Average Score}}_{(Process\ Capabilities)} = (Score\ \Sigma_{i=1}^{n} P_i)/n$$
$$= (4 + 3)/2$$
$$= 7/2 = 3.5$$

$$\textbf{\textit{Overall Sustainability Index Value}} = Average\ scores/n = (4.67 + 3.67 + 3.5)/3$$
$$= 11.84/3 = 3.95 < 2.5\ (\text{minimum score to qualify})$$

For the present case, the sustainability index value of the microwave sintering process is close to 4, which shows the process is highly sustainable. This is based on the simple method of weighted averages but can provide significant information regarding process sustainability. Scores can also be collected through surveys from experienced user groups for better predictions on sustainability index values. The different parameter categories can be added per the requirements, and various parameters affecting the overall manufacturing process can be used to track its performance. For any process to qualify as sustainable, it should score at least 2.5 out of 5 considering all the major parameters and comparisons to conventional process values. This score can also provide insights on areas of improvement for the particular process and critical areas hampering the sustainability of process. Table 1.2 summarizes the scores of various parameter categories and the sustainability index.

This case study reveals that the sustainability of a process is based on the sustainability of many parameters, which needs to be addressed before actually calculating the sustainability for any process. Benchmarking needs to be done against the parameter values so that a real picture of the process can be obtained. Figure 1.2 shows a methodology chart for calculating the sustainability value for any process.

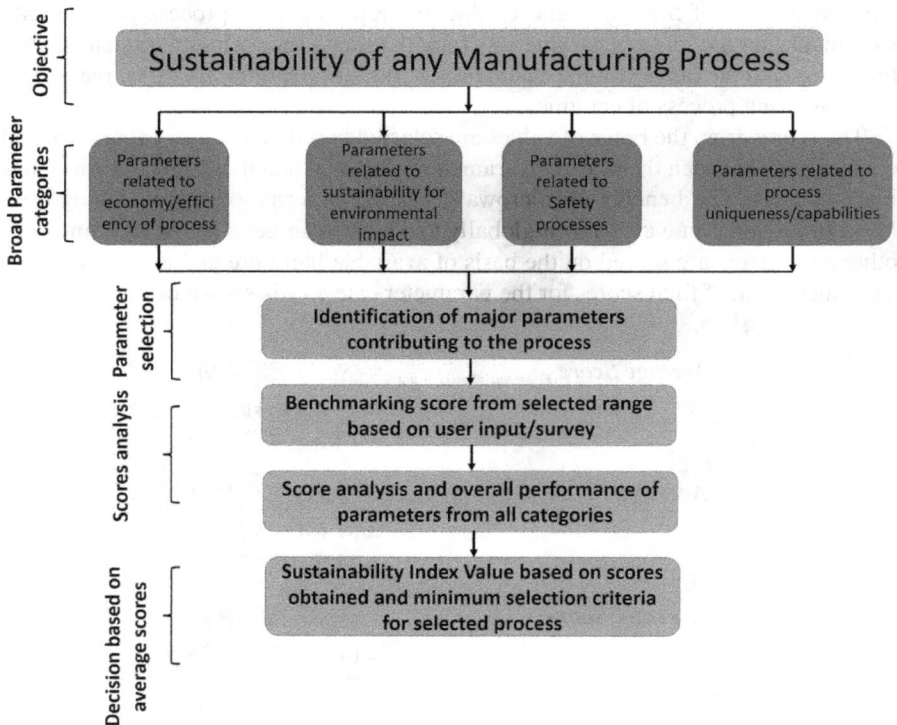

FIGURE 1.2   Proposed methodology for calculating the sustainability index value.

1.3   CONCLUSION

There is a global need for sustainable manufacturing processes that can help industries grow further. Sustainability itself is a wide term that can include many parameters for process sustainability. The frameworks available in the literature are too
complex to work on and implement. The present work reports a simple method to
analyze any manufacturing process/material processing technology for sustainability. The method involves five steps to find the sustainability index:

- First, identify the major parameter categories for any process that directly
  affects sustainability.
- For every category, list the major parameters.
- Compare these parameters to conventional available processes or get input
  from specialist user groups through surveys.
- Provide a general score based on the inputs received. Calculate the average
  score for each parameter category.
- Calculate the total average score of all categories to find the sustainability
  index value. A minimum target is to be set at, say, 2.5 (50%) on average for
  all parameter categories, and if the sustainability index value is more than
  2.5, the process can be considered sustainable.

Further, the data from this analysis can highlight the category in which the process is underperforming or sustainability is hampered. This method can provide a
simple way to gain insights on the sustainability issues of any manufacturing process.

REFERENCES

Abualfaraa, W., Salonitis, K., Al-Ashaab, A., & Ala'raj, M. (2020). Lean-green manufacturing
   practices and their link with sustainability: A critical review. *Sustainability*, *12*(3), 981.
Arumugam, V., Kannabiran, G., & Vinodh, S. (2022). Impact of technical and social lean
   practices on SMEs' performance in automobile industry: A structural equation modelling (SEM) analysis. *Total Quality Management & Business Excellence*, *33*(1–2), 28–54.
Böhringer, C., & Jochem, P. E. P. (2007). Measuring the immeasurable—A survey of sustainability indices. *Ecological Economics*, *63*(1), 1–8.
Cai, W., Lai, K.-H., Liu, C., Wei, F., Ma, M., Jia, S., Jiang, Z., & Lv, L. (2019). Promoting
   sustainability of manufacturing industry through the lean energy-saving and emission-
   reduction strategy. *Science of the Total Environment*, *665*, 23–32.
Comm, C. L., & Mathaisel, D. F. X. (2005). A case study in applying lean sustainability
   concepts to universities. *International Journal of Sustainability in Higher Education*,
   6(2), 134–146.
Hallam, C. R. A., & Contreras, C. (2018). Lean healthcare: Scale, scope and sustainability.
   International Journal of Health Care Quality Assurance, *31*(7), 684–696.
Herrmann, C., Schmidt, C., Kurle, D., Blume, S., & Thiede, S. (2014). Sustainability in manufacturing and factories of the future. *International Journal of Precision Engineering
   and Manufacturing—Green Technology*, *1*(4), 283–292.
Lucato, W. C., Santos, J. C. da S., & Pacchini, A. P. T. (2018). Measuring the sustainability of
   a manufacturing process: A conceptual framework. *Sustainability (Switzerland)*, *10*(1).
   http://doi.org/10.3390/su10010081

Martínez León, H. C., & Calvo-Amodio, J. (2017). Towards lean for sustainability: Understanding the interrelationships between lean and sustainability from a systems thinking perspective. *Journal of Cleaner Production*, *142*, 4384–4402.

Pathania, A., Singh, S., Gupta, D., & Jain, V. (2015). Development and analysis of tribological behavior of microwave processed EWAC + 20% WC10Co2Ni composite cladding on mild steel substrate. *Journal of Manufacturing Processes*, *20*, 79–87.

Singh, S., Gupta, D., Jain, V., & Sharma, A. K. (2015). Microwave processing of materials and applications in manufacturing industries: A review. *Materials and Manufacturing Processes*, *30*(1), 1–29.

2 Microwave Irradiation Manufacturing of Polymer Composites
A Sustainable Manufacturing Technique

Gaurav Arora

CONTENTS

ABBREVIATIONS

AFM	Atomic force microscopy
AP	Applicator
ASTM	American Society for Testing and Materials
CNTs	Carbon nano-tubes
CuO	Copper oxide
DBTM	Dibutyltindimethoxide
DSC	Differential scanning calorimetry
FT-IR	Fourier transform infrared spectroscopy
HA	Hydroxyapetite
HDPE	High-density polyethylene
MAC	Microwave-absorbing constituent
MHz	Megahertz
MIM	Microwave irradiation manufacturing
MMA	Methyl methacrylate
MWCNTs	Multi-walled carbon tubes
PANI	Polyaniline
PCS	Photon correlation spectroscopy
PE	Polyethylene
PLA	Poly-lactic acid

DOI: 10.1201/9781003269298-2

PLLA	Poly-l-lactic acid
PMC	Polymer matrix composites
PP	Polypropylene
SEM	Scanning electron microscopy
SR	Source
SWCNTs	Single-walled carbon nano-tubes
TEM	Transmission electron microscopy
TGA	Thermo-gravimetric analysis
TL	Transmission lines
VSM	Vibrating-sample magnetometer
XRD	X-ray diffraction

2.1 INTRODUCTION

Electromagnetic waves in the frequency range 1 to 300 GHz and wavelength range 1 mm to 1 m are known as microwaves (Taylor et al., 2015; Thostenson & Chou, 1999). The Federal Communications Commission has reserved two frequencies, 0.915 and 2.45 GHz, to be used for industrial, medical, and science purposes (Taylor et al., 2015; Thostenson & Chou, 1999). In 1993, microwave furnaces operating between 0.9 and 18 GHz were developed for processing material (Taylor et al., 2015; Thostenson & Chou, 1999), shown in Figure 2.1.

Microwave furnaces have three main operating components: the source (SR), transmission lines (TL), and applicator (AP), shown in Figure 2.2. The function of the source is to generate electromagnetic waves, which are further delivered to

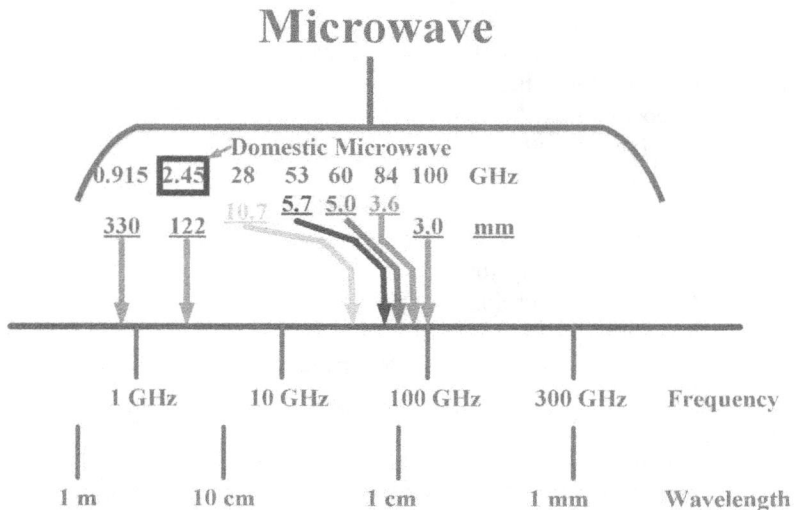

FIGURE 2.1   Frequency and wavelength spectrum of microwaves.

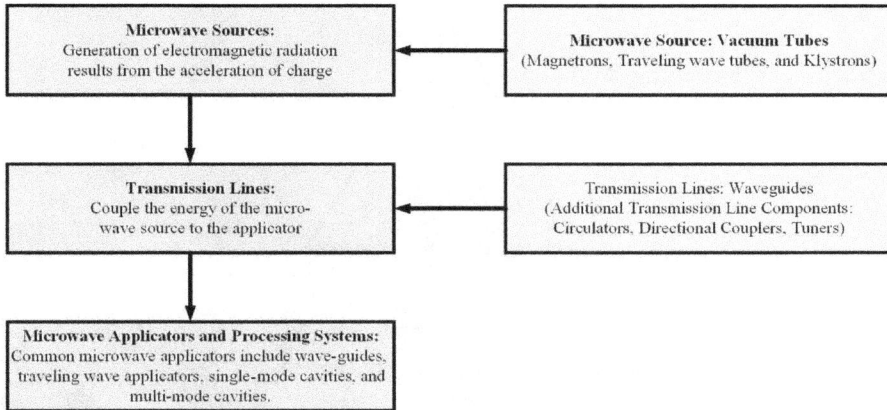

FIGURE 2.2 Flow chart of microwave processing unit.

the applicator by the transmission lines (Taylor et al., 2015; Thostenson & Chou, 1999).

The theoretical study of these components can be done by the appropriate boundary conditions and Maxwell equations as follows (Thostenson & Chou, 1999):

$$\nabla \times \mathbf{E} = \frac{\partial \mathbf{B}}{\partial t}, \quad \nabla \cdot \mathbf{B} = 0; \tag{1}$$

$$\nabla \times \mathbf{H} = \frac{\partial \mathbf{D}}{\partial t} + \mathbf{I}, \quad \nabla \cdot \mathbf{D} = \rho; \tag{2}$$

where $\mathbf{E}$, $\mathbf{B}$, $\mathbf{H}$, $\mathbf{D}$, $\mathbf{I}$, and ρ are the electric field vector, magnetic flux density vector, magnetic field vector, electric flux density vector, current density vector, and charge density, respectively. An efficient microwave oven system can only be designed by combining the components (SR, TL, and AP) with proper knowledge and understanding of electromagnetic theory for the processing of materials (Thostenson & Chou, 1999).

MIM is a sustainable manufacturing technique utilized for processing metals and non-metals. The evolution of processing materials is represented in Figure 2.3. MIM is the cheapest, fastest, easiest, and greenest non-conventional manufacturing technique. Products manufactured using MIM can be used in different applications like aerospace, automobile, commercial, and so on.

Microwave irradiation has been used for different studies by various researchers, such as recovery of carbon fibers for reuse in high-quality applications (Lester et al., 2004), the study of interfacial properties and microstructural evaluation of nanoparticle-reinforced composites (Dong et al., 2008), preparation of thermos-responsive nanoparticles by emulsion polymerization (Huang et al., 2008), development of glass polymer composites in different magnetic fields (S. Li et al., 2010), studying the effect of kinetics of polymerization of composites subjected to conventional and microwave

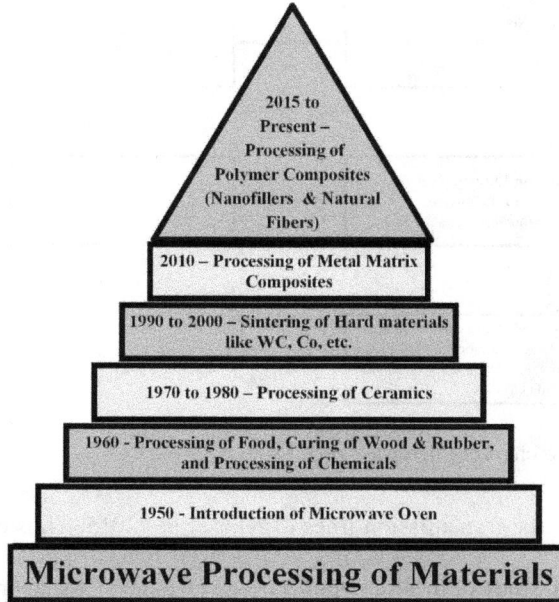

FIGURE 2.3   Evolution of microwave processing of materials.

irradiation (Spasojevi, 2013), and studying the effect of clay content on in-situ fabricated microwave irradiation polymer nanocomposites (Singla et al., 2014).

It is evident from the literature that MIM has had a great impact on the processing of composites. From an MIM perspective, composites fall under the category of mixed absorbing materials; therefore, the final properties (electrical, thermal, mechanical, thermo-mechanical, etc.) of composites depend on the constituents. The coupling of microwaves in composites only occurs with the constituent of a high dielectric loss factor. Thus, the presence of such a constituent is a necessary condition in composites produced by the MIM technique. Absorption of microwaves by this constituent is the key factor in conducting heat from the inside out in the composites. The local heat produced by the absorption and then conduction to nearby constituents is the uniform heating phenomenon that affects the overall properties of the composites compared to conventional heating techniques.

However, the second factor that affects the overall properties of composites is the volume concentration of the microwave absorbing constituent (MAC). In the literature, many composites with MACs, like C powders, BNNTs, CNTs, Ni, Al, and Cu, have been studied, and it has been reported that these MACs also control the conductive properties (Arora et al., n.d., 2019, 2021; Arora & Pathak, 2020, 2021; Moulart et al., 2004; Padmanabhan et al., 2020; Pramila Devi et al., 2012; Skorokhod, 1995). But it has also been reported that optimum volume concentration is also necessary in order to reduce the inferior properties of composites due to under- or overheating of the matrix constituent.

TABLE 2.1

Comparison of Dielectric Loss Factor of Polymers and Fillers

Materials	Dielectric Loss Factor
Polymers (at room temperature)	0.008–1.20
Natural fillers (at room temperature)	0.0010–0.0068
Synthetic fillers (at room temperature)	1.4–3.75

Source: Mishra & Sharma, 2016.

MIM of polymer composites mainly depends on the frequency of processing, dipole structure, temperature, and characteristic properties of the constituents (polymer and filler) (Mishra & Sharma, 2016). The dominant selective heating of MAC (filler) over the matrix results in the effective manufacturing of polymer composites. The literature on polymer composites reinforced with synthetic and natural fillers is vast. It had been reported that in the case of natural filler reinforced composites, the polymer matrix absorbs microwaves due to a high dielectric loss factor compared to natural fillers. On the other hand, synthetic fillers dominate processing using MIM technique due to a high dielectric loss factor compared to a polymer matrix. Table 2.1 represent the dielectric loss factor of polymers and fillers.

2.2 MIM: PROCESSING OF POLYMER COMPOSITES

The literature on MIM technique is vast, and it has been reported that MIM can easily join, clad, drill, and cast metals and non-metals uniformly and with minimum microwave exposure (Agrawal, 2006; Benítez et al., 2007; Bhoi et al., 2019; Boey & Lee, 1990; Fotiou et al., 2013; Gamit et al., 2017; Gupta et al., 2012; Gupta & Sharma, 2014; Huang et al., 2008; Johnsona et al., 1997; Y. Li et al., 2019; Lin et al., 2009; Oh et al., 2020; Poyraz et al., 2015; M. K. Singh, Zafar, et al., 2019; M. K. Singh & Zafar, 2018; S. Singh et al., 2018; Spasojevi, 2013; Tamang & Aravindan, 2017; Wang et al., 2007; Wise & Froment, 2001; Xie et al., 2011; Zabihi et al., 2020). Therefore, it can be concluded that MIM is a fast, efficient, and reliable manufacturing technique. Also, it has been reported that filler materials can be recovered by microwave irradiation by diminishing the adverse environmental impacts while preserving energy and time, thus making it a sustainable manufacturing method (Lester et al., 2004; Zabihi et al., 2020).

Microwave-assisted synthesis of Fe_3O_4/poly(styrene-co-acrylamide) with emulsion polymerization was described by Huang et al. (2008). The different properties of the prepared nanoparticles were studied by different characterizations like FT-IR, TEM, TGA, XRD, and PCS. Furthermore, with the help of hysteresis curves determined at room temperature, the magnetic properties were measured. The study confirms the diffusion of Fe_3O_4 particles into microwave-assisted synthesized Fe_3O_4/poly(styrene-co-acrylamide) nanoparticles because a consistent pattern was observed. The C-H bonding of the benzene ring was characterized by the FT-IR

TABLE 2.2

Recent Literature on MIM Processing of Polymer Composites (continued)

Composite Type	Characterizations	MIM Technique	Reference
Fe_3O_4/poly(styrene-co-acrylamide)	Magnetization, TGA, TEM, XRD	Microwave-assisted synthesis with emulsion polymerization	(Huang et al., 2008)
$Cu_{50}Zr_{45}Al_5$ metallic glass/polyphenylene sulfide	Heating behavior, XRD	Microwave irradiation with separated H-field and applied pressure of 5 MPa	(Li et al., 2010)
Poly(methyl methacrylate) composites	Kinetic parameters	Microwave isothermal heating	(Spasojevi, 2013)
Poly (lactic acid)/clay nanocomposites	SEM, TEM, XRD, TGA	Microwave-assisted in situ ring opening polymerization	(Singla et al., 2014)
CuO nanosheets/epoxy	SEM, TEM, XRD, Raman spectra, mechanical properties (tensile)	Microwave-assisted route	(Nahas et al., 2016)
Fe_3O_4/Poly(styrene-divinylbenzene-acrylic acid) polymeric magnetic composites	XRD, TGA, SEM, TEM, magnetization	Microwave-irradiated emulsion polymerization	(Jaiswal et al., 2017)
Carbon fiber composites	TGA, SEM, Raman spectra, mechanical properties (tensile)	Microwave-assisted chemical method	(Zabihi et al., 2020)
Polymer/NaLa $(MoO_4)_2$-G-PPy nanocomposites	TEM, UV–visible spectroscopy, Raman spectra	Microwave-assisted synthesis	(Oh et al., 2020)
Salecan/PAPTAC/Fe3O4@TA composite hydrogels	FT-IR, XRD, TGA, SEM, compressive, magnetic and rheological properties	Microwave-assisted polymerization	(Hu et al., 2018)
Nanocomposite nylon-6/graphene	FT-IR, TGA, SEM	Microwave-assisted polymerization	(González-Morones et al., 2018)
T700/QY9611 continuous carbon fiber-reinforced bismaleimide prepreg	Microwave transmissivity, infrared thermal imaging, computational analysis	Vertical penetrating microwave (VPM)	(Li et al., 2019)

TABLE 2.2
Recent Literature on MIM Processing of Polymer Composites

Poly (L-lactide) (PLLA)/nano-hydroxyapatite (HA) composites	Mechanical properties (compression and flexural strength), SEM	Microwave curing	(Kumar et al., 2019; Singh, Zafar, et al., 2019)
Poly (L-lactide) (PLLA)/coir composites	Mechanical properties (uniaxial tension and flexural tests), SEM	Microwave processing	(Singh, Verma, et al., 2019)
Polypropylene/kenaf, Polypropylene/jute, Polyethylene/kenaf, Polyethylene/jute composites	Mechanical properties (uniaxial tension, impact tests), SEM, XRD	Microwave processing	(Singh, Verma, & Zafar, 2020; Singh & Zafar, 2019, 2020a, 2021; Tewari et al., 2020)
Polythylene/kenaf/jute hybrid composites and pinecone/recycled polyethylene composites	Mechanical properties (uniaxial tension, impact tests, flexural tests), SEM	Microwave-assisted compression molding	(Kumar et al., 2022; M. Singh & Zafar, 2020b; Tewari et al., 2021)
Polyethylene/CNTs and Polypropylene/CNTs nanocomposites	Mechanical properties (uniaxial tension and fracture toughness, nano-indentation), SEM, XRD	Microwave processing	(Arora et al., 2019; Arora & Pathak, 2020, 2021; Pundhir et al., 2019)

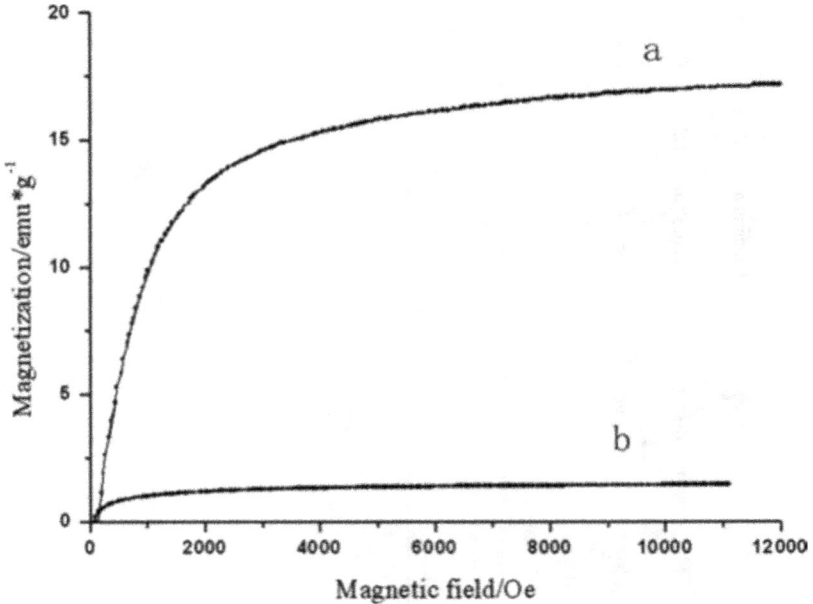

FIGURE 2.4 Magnetization versus magnetic field for (a) magnetic fluid; (b) magnetic Fe_3O_4/poly(styrene-co-acrylamide) microspheres (Huang et al., 2008).

examination for poly(St–AAm) nanocomposite in the band range of 3100–3000 cm⁻¹ and for Fe_3O_4/poly(St–AAm) nanoparticles in the band range of 2000–1668 cm⁻¹.

Magnetic materials display magnetic hysteresis loops as one of their main characteristics. An external magnetic field determines whether magnetic materials can respond to the field. VSM was used to examine the magnetic properties of a magnetic fluid and magnetic Fe_3O_4/poly(St–AAm) nanoparticles. The magnetization curves for the magnetic fluid and the magnetic Fe_3O_4/poly (St–AAm) microspheres at room temperature are shown in Figure 2.4. There were approximately 17.2 and 1.45 emu/g saturation magnetizations for the magnetic fluid and magnetic Fe_3O_4/poly(St–AAm) nanoparticles, respectively. Magnetic fluid has a saturation magnetization that is significantly less than bulk magnetite, which has a saturation magnetization of 84 emu/g.

By microwave processing the constituents in a separate H-field under an applied pressure of 5 MPa, a novel $Cu_{50}Zr_{45}Al_5$ metallic glass/polyphenylene sulfide (PPS) composite with high relative densities was fabricated (S. Li et al., 2010). The study used a single-mode microwave apparatus of 915 MHz. X-ray diffractometry in reflection with monochromatic Cu–Kα radiation was used to determine the structures of the initial powders and sintered composites. SEM was used to characterize the microstructures of the sintered samples and powder metallic alloy.

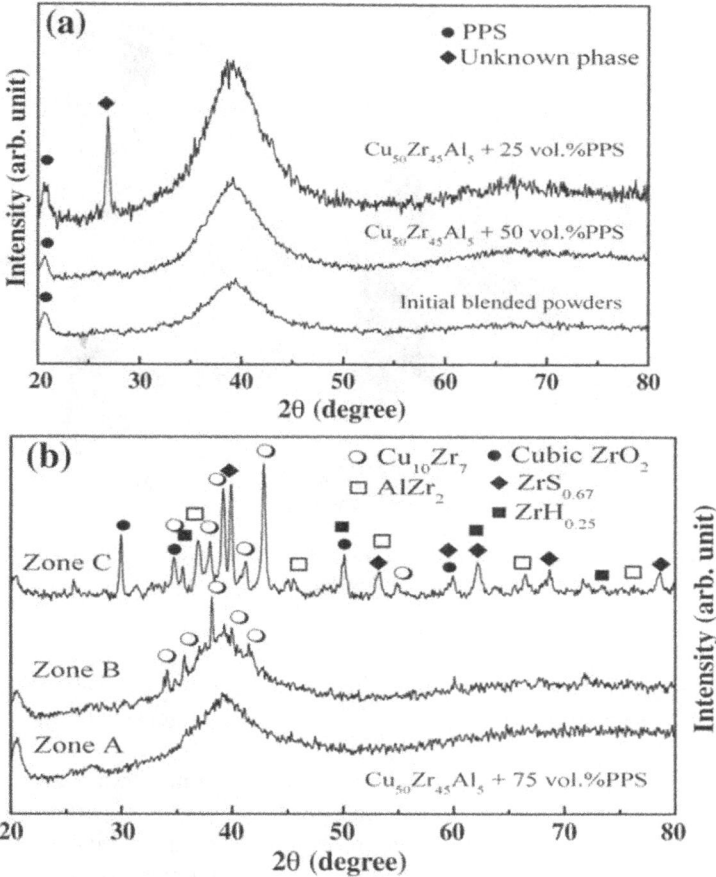

FIGURE 2.5 (a) XRD patterns of the initial blended powders and composites with 25 and 50 vol.% PPS powder heated by microwaves under a flowing nitrogen gas; (b) XRD patterns of different zones of the composite with 75 vol.% PPS (S. Li et al., 2010).

From the XRD examination (Figure 2.5), it is evident that the composite with 75 vol.% PPS was induced by microwave heating in a separated H-field to have a gradient structure in contrast to the composites with 25 and 50 vol.% PPS. The thermal gradient was confirmed by using the structural change mentioned as zones A, B, and C. Zone A had a structure similar to the powder. Zone C had the porous structure depicted in the center. The cubic ZrO_2, $ZrS_{0.67}$, and $ZrH_{0.25}$ phases were detected due to the decomposition of PPS during microwave irradiation. PPS powder was found to be cured in zone B using microwaves, and retention of the metallic glassy phase of $Cu_{50}Zr_{45}Al_5$ was observed. The dielectric properties of the constituents produced the gradient structure when processed using the microwave in the composite. SEM examination (Figure 2.6) showed rapid and excellent

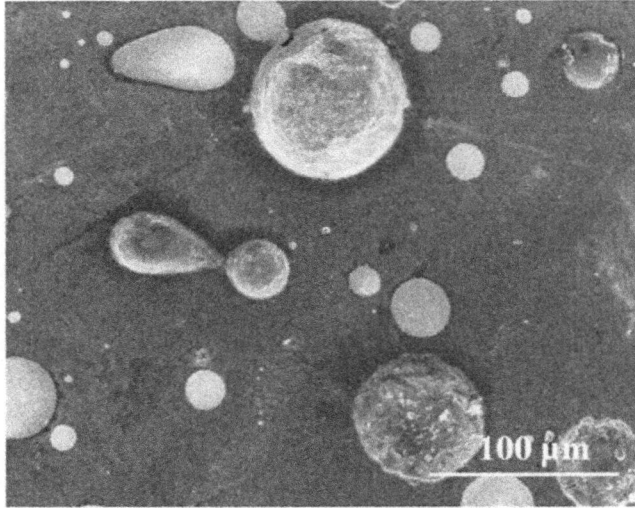

FIGURE 2.6   SEM micrograph of the polished cross-section of the composite with 75 vol.% PPS powder heated by microwaves under a flowing nitrogen gas (S. Li et al., 2010).

consolidation of PPS powders through microwave heating in zone B. The wetting seemed good, with no obvious cracking observed at the interface between metallic glassy particles and PPS.

Researchers explored the effects of conventional and microwave heating on the kinetics of isothermal polymerization in MMA composite materials (Spasojevi, 2013). Microwave heating results in lower polymerization kinetic parameters than conventional heating, as illustrated in Table 2.3, characterized by Equation 3 (Spasojevi, 2013).

$$\left(\frac{d\alpha}{dt}\right)_{max} = \mathrm{Ln}[A \cdot f(\alpha_{max})] - \frac{E_a}{RT} \tag{3}$$

The increase in the energy of vibration of C-O in the methyl methacrylate molecule accompanied by a decrease in the anharmonicity factor resulted in a decrease in E_a and $\mathrm{Ln}(A)$ with microwave heating in comparison to conventional heating. In both heating modes, MMA composite material polymerizes by a kinetic mechanism. The increase in the polymerization rate under microwave heating compared to conventional heating cannot be attributed to hot spots, overheating, or selective heating in the reaction system.

A first report of the significant effect of clay mineral on the in situ polymerization of lactide using microwave irradiation was given by Singla et al. (2014). Using microwave heating and SnOct$_2$ and DBTM as initiators, PLA/clay nanocomposites were prepared in situ. Three different clays—Clay type 1: natural unmodified montmorillonite, Clay type 2: montmorillonite modified with di-methyl dehydrogenated

TABLE 2.3
Kinetic Parameters and Rate of Process for Conventional and Microwave Heating of the Composite at Different Temperatures

Temperature, (K)	Conventional Heating		Microwave Heating	
	Max Rate of the Process, $\left(\dfrac{d\alpha}{dt}\right)_{max}$ (min^{-1})	Kinetic Parameters	Max Rate of the Process, $\left(\dfrac{d\alpha}{dt}\right)_{max}$ (min^{-1})	Kinetic Parameters
343	0.0127	Activation energy, $E_a = 86 \pm 4$ kJ mol^{-1}	0.0583	Activation energy, $E_a = 72 \pm 5$ kJ mol^{-1}
353	0.0282	Pre-exponential factor, $Ln(A) = 24.8 \pm 1.5$	0.1094	Pre-exponential factor, $Ln(A) = 22.2 \pm 2.0$
363	0.0679		0.2350	

Source: Spasojevi, 2013.

tallow quaternary ammonium cations, and Clay type 3: montmorillonite modified with methyl tallow bis-2-hydroxyethyl quaternary ammonium cations—were used in this study (Singla et al., 2014). By applying microwave irradiation, the polymerization process is accelerated and exfoliated structures are formed. As a result of the study, microwave irradiation in situ polymerization seems to be a resourceful, cost-efficient, and ecologically friendly method for preparing clay-based nanocomposites.

A study was conducted on synthesis of CuO nanosheets fabricated by using the microwave technique and used as fillers in the thermosetting matrix, that is, epoxy (Nahas et al., 2016). The weight fraction of the nanosheets used varied from 0.1% to 2.0% wt. The mechanical properties of the nanocomposites were investigated using the uniaxial tensile test. It was observed that with the addition of 0.5% wt. of CuO nanosheets to epoxy, the stiffness, stress at fracture, strain at fracture, and elongation at fracture improved to 63%, 27%, 40%, and 34%, respectively, compared to neat epoxy. With higher concentration of CuO nanosheets (1% and 2% wt.), it was found that the mechanical properties were reduced due to agglomeration. The tensile stress–strain curve for the different concentrations of CuO nanosheets is represented in Figure 2.7. Apart from this, the degassing effect to remove voids also showed significant improvement in the mechanical properties. These composites can be used for high-stiffness products.

Polymeric magnetic composites were synthesized with the help of microwave-irradiated emulsion polymerization (Jaiswal et al., 2017). The synthesized composites were studied for the effect of reaction time and amount of reactive reagents on their magnetic and thermal properties along with morphology and structure. Fe_3O_4 nanoparticles synthesized at 300 K exhibit superparamagnetic behavior with a saturation magnetization of 67.5 emu/g. The composites with poly(styrene-divinyl-benzene-acrylic acid) and Fe_3O_4 nanoparticles, that is, PMC-10 (acrylic acid concentration 1.5 g) and PMC-11 (acrylic acid concentration 2.0 g) loaded with 24%

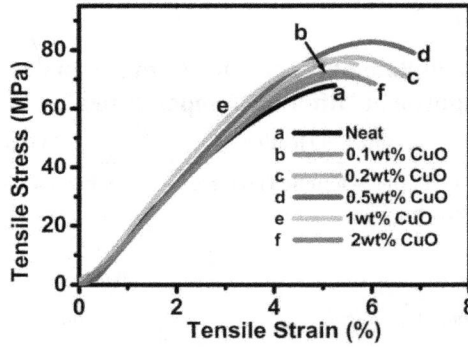

FIGURE 2.7 Stress-strain curve of CuO nanosheets reinforced epoxy composites (Nahas et al., 2016).

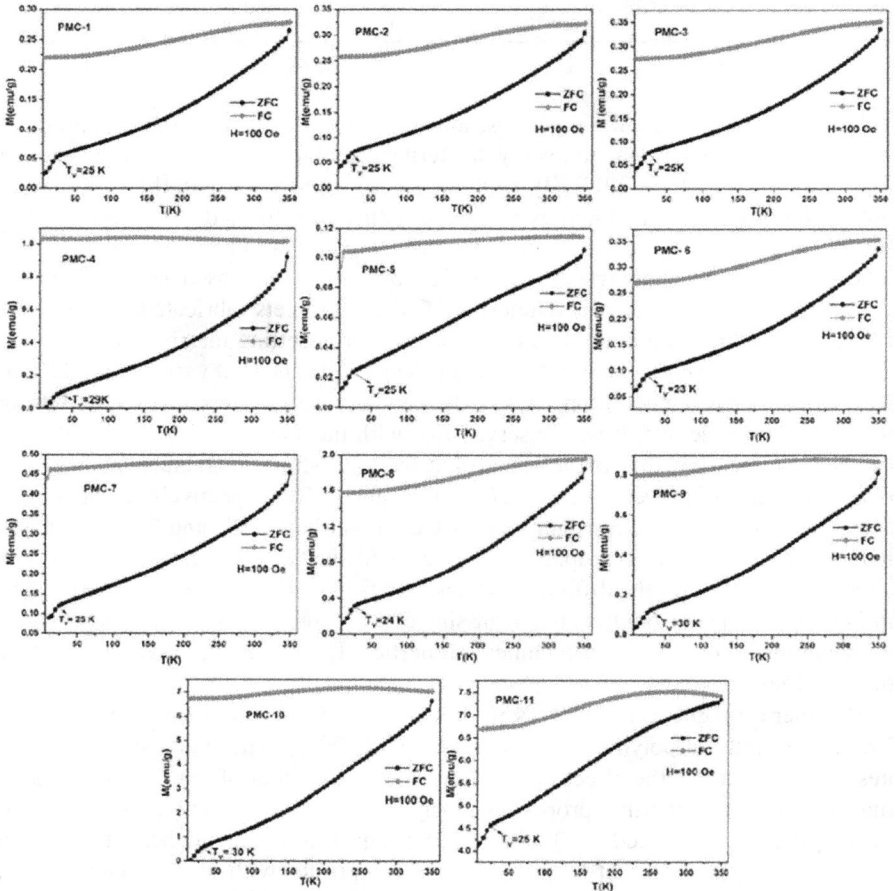

FIGURE 2.8 Temperature-dependent magnetization of PMC-1 to PMC-8 with 1.0 g acrylic acid, PMC-9 with 0.5 g acrylic acid, PMC-10 with 1.5 g acrylic acid and PMC-11 with 2.0 g acrylic acid (Jaiswal et al., 2017).

magnetic nanoparticles, had a maximum saturation magnetization of 22.5 and 18.7 emu/g, respectively. These microwave-irradiated composites can be favorable materials for magnetic sensors and biomedical applications.

Porous biocomposites of PLLA/HA (poly-l-lactic acid reinforced hydroxyapatite) and natural fiber composites (kenaf/coir/pinecone/jute reinforced polymer) have been successfully developed using a modern named technology, microwave-assisted compression molding (Kumar et al., 2019; Kumar Singh et al., 2021; Pundhir et al., 2019; M. K. Singh, Verma, & Zafar, 2020; M. K. Singh, Verma, et al., 2019; M. K. Singh, Verma, Pundhir, et al., 2020; M. K. Singh & Zafar, 2018, 2019, 2020b; Tewari et al., 2020, 2021). In a study done on the fabrication of porous biocomposites, that is, a poly-l-lactic acid reinforced hydroxyapatite scaffold, it was observed that microwave assistance resulted in nearly 22% porosity in the structure compared to theoretical

TABLE 2.4

Mechanical Properties of Microwave-Assisted Compression Molding Natural Fiber Composites

Natural Fiber Composite	Tensile Strength (MPa)	Fracture Strain (%)	Flexural Strength (MPa)	Impact Strength (kJ/m²)	Hardness (Shore D)
Coir/HDPE	29.5 with 10% coir	6 with 10% coir	32.14 with 15% coir	NR	NR
Coir/PLLA	18 with 30% coir	NR	48 with 30% coir	NR	NR
Pinecone/recycled HDPE	20.01 with 20% pinecone	6 with 20% coir	20.98 with 20% pinecone	16 with 20% pinecone	65 with 50% pinecone
PP/kenaf	49.75 with 15% kenaf	NR	NR	24 with 15% kenaf	NR
PP/jute	44 with 15% jute	NR	NR	18 with 15% jute	NR
PE/kenaf	46.5 with 15% kenaf	NR	NR	22 with 15% jute	NR
PE/jute	46.3 with 15% jute	NR	NR	20 with 15% kenaf	NR
PE/jute (three-layer)	49.75	6	50.13	21.1	53.7
PE/kenaf (three-layer)	60	8.5	54.33	31.8	58
PE/(jute/kenaf/jute)	60	5.5	55.64	23.3	54.5
PE/(kenaf/jute/kenaf)	54	4.8	40.38	27.7	57.1

Note: NR = not reported.

Source: Kumar et al., 2019; Kumar Singh et al., 2021; Pundhir et al., 2019; M. K. Singh, Verma, & Zafar, 2020; M. K. Singh, Verma, et al., 2019; M. K. Singh, Verma, Pundhir, et al., 2020; M. K. Singh & Zafar, 2018, 2019, 2020b; Tewari et al., 2020, 2021.

21–23% porosity. The compressive and flexural strength of the composites was found to be 11.17 and 95.4 MPa for 10% HA and neat PLLA, respectively.

Table 2.4 represents the mechanical properties of microwave-assisted compression-molded natural fiber-reinforced composites. It can be concluded from Table 2.4 that chopped coir/HDPE and coir/PLLA composites show maximum flexural strength compared to tensile strength. Also, the concentration (%) of coir in the composites was 10% and 30%, respectively (M. K. Singh, Verma, et al., 2019; M. K. Singh, Verma, Pundhir, et al., 2020; M. K. Singh & Zafar, 2018). Forest waste, pinecones, have been used as filler in recycled PE of about 20% concentration. It has been reported that the tensile strength and flexural strength of pinecone/PE composites are 20.01 and 20.98 MPa, respectively. The maximum impact strength and hardness (shore D) of the same composite were reported as 16 kJ/m^2 and 65 with 20% and 50% concentrations of pinecone, respectively.

Much research on the use of jute and kenaf as natural fillers in thermoplastic polymers for low-load applications such as lamina and laminate has also been reported (M. K. Singh, Verma, & Zafar, 2020; M. K. Singh & Zafar, 2019, 2020a, 2020b, 2021; Tewari et al., 2020). As reported in Table 2.4, jute and kenaf have been used as filler with PE and PP for the performance evaluation of the newly developed microwave-assisted compression-molding technique. Mechanical properties like tensile strength, fracture strain, flexural strength, impact strength, and hardness were the evaluating parameters for the performance of the newly developed technique.

The reported PE/jute, PE/kenaf, PP/jute, PP/kenaf, PE/jute (three-layer), PE/kenaf (three-layer), PE/(jute/kenaf/jute), and PE/(kenaf/jute/kenaf) composites were uniaxially tested for tensile properties with 46.3, 46.5, 44, 49.75, 49.75, 60, 60, and 54 MPa ultimate tensile strength, respectively. Good interfacial bonding in the composites resulted in the tensile fracture of the outer surface in PE/kenaf (three-layer) and PE/(kenaf/jute/kenaf) hybrid composites. However, interlaminar fracture was found in PE/jute (three-layer) and PE/(jute/kenaf/jute) hybrid composites. Another reason for the higher tensile strength of PE/kenaf (three-layer) was less cracking and more elongation compared to jute, which therefore acted as a load transfer medium during the testing.

The fracture strain of 8.5% is found to be the maximum for PE/kenaf (three-layer) compared to PE/jute (three-layer) natural fiber composites because of imperfect interfacial bonding examined by SEM for PE/jute (three-layer) (M. K. Singh & Zafar, 2020b). The flexural strength of 55.64 MPa has been found to be the maximum for PE/(jute/kenaf/jute) hybrid composites due to fiber bridging in this composite compared to PE/(kenaf/jute/kenaf), examined through SEM (M. K. Singh & Zafar, 2020b). The impact strength of 31.8 kJ/m^2 has been found to be the maximum for PE/kenaf (three-layer) composites. The decrease in the impact strength of pinecone/recycled HDPE can be defined on the basis of induced microcracks in the matrix due to moisture present in the pinecone (Tewari et al., 2021). This resulted in lower interfacial bonding and also lower impact energy (Tewari et al., 2021).

The maximum hardness (shore D) values of 65 and 57.1 were found for pinecone/recycled HDPE and PE/(kenaf/jute/kenaf) hybrid composites, respectively. An increment of 8% in PE/kenaf (three-layer) was found compared to PE/jute (three-layer) composites. In the other hybrid composites, the decrease in hardness is due to an increase in void content.

The method to increase the interfacial adhesion between the carbon fiber and the polymer matrix using the microwave irradiation process before the curing of prepags using low energy beam curing resulted in improved mechanical properties (Zhang et al., 2020). The study was conducted on PAN-based T700 carbon fiber with epoxy resin. The microwave temperature irradiated on the prepag and epoxy was between 5 °C and 150 °C. The carbon fiber surface condition was examined by SEM and AFM. Mechanical properties like interlaminar shear strength and tensile strength of the composites were examined following ASTM standards 2334 and 3039, respectively (Zhang et al., 2020).

Physical entangling of resin on the carbon fiber surface has been reported with an increase in microwave processing time from 0 to 180 s. This has been characterized by AFM such as the roughness increasing from 11.3 to 35.3 nm from 0 to 180 s exposure time (Zhang et al., 2020). Thus, this may improve the interfacial shear strength by enhancing the mechanical keying between the constituents.

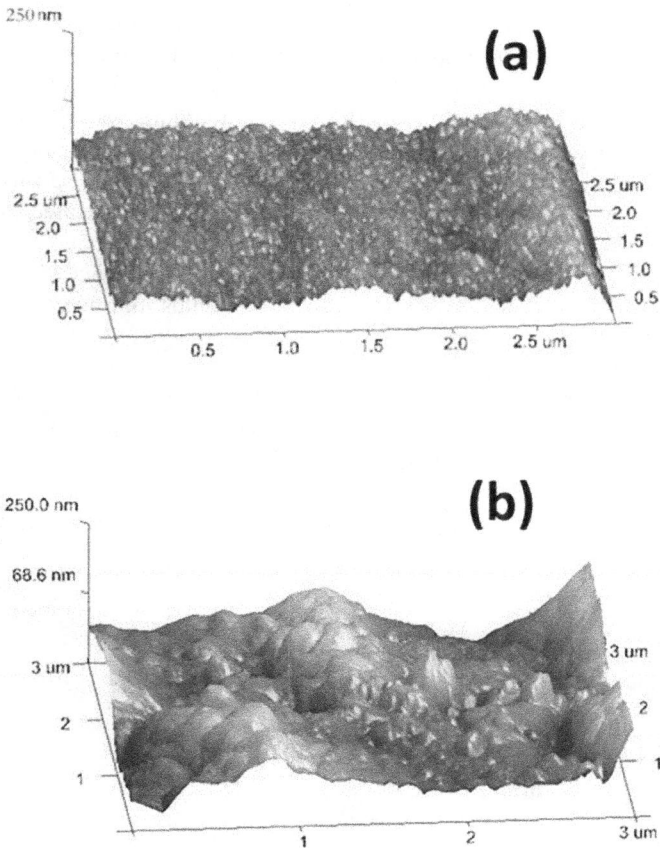

FIGURE 2.9   AFM of carbon fibers extracted from microwave irradiated composite at (a) 0 s and (b) 180 s (Zhang et al., 2020).

The interfacial shear strength increased from 34.93 to 45.95 MPa with an increase in microwave exposure time from 0 to 180 s. In the case of the laminate, the maximum interlaminar shear strength of 56.5 MPa was observed at 90 s exposure time. The resin-poor areas are the main reason behind the drop of interlaminar shear strength at 180 s exposure time. A maximum and minimum tensile strength of 1421 and 1225 MPa were found in the composites exposed to 90 and 180 s, respectively. This indicates that the increase in interfacial property is at the expense of tensile strength. Therefore, the study suggests not irradiating longer than 90 s (Zhang et al., 2020).

Another study on the microwave curing of carbon fiber polymer composites resulted in successful curing of 2.3-mm-thick composite by penetrating the microwaves vertically, as shown in Figure 2.10 (Li et al., 2019). The study was done on unidirectional and multidirectional carbon fiber-reinforced composites. The conclusion of the study was that the vertical microwave penetration technique is successful only for multidirectional carbon fiber-reinforced composites.

Another study on carbon fiber-reinforced polymer woven prepreg composites curing was also reported (Joshi & Bhudolia, 2014). In this study, a comparison of thermal, microwave, and combined thermal and microwave technique was the focus. Out of these three, the combined thermal and microwave technique was found to be the effective method for saving time and energy without compromising the mechanical properties. It was also concluded that this combined process is 2.5 times faster compared to the autoclave process. Eight-layer laminates were fabricated for tensile and flexural testing. A 1.7% and 5.5% improvement in the tensile strength and tensile modulus of the combined process were observed compared to the autoclave process (Joshi & Bhudolia, 2014). The flexural strength and modulus improved to 2.5% and 8.6% in combined process compared to the autoclave process (Joshi & Bhudolia, 2014).

The potential of curing polymer composites at 915 MHz was evaluated in comparison to microwave curing at 2450 MHz and thermal curing (Rao et al., 2020). The interlaminar shear strength and flexural strength were the evaluating criteria. The

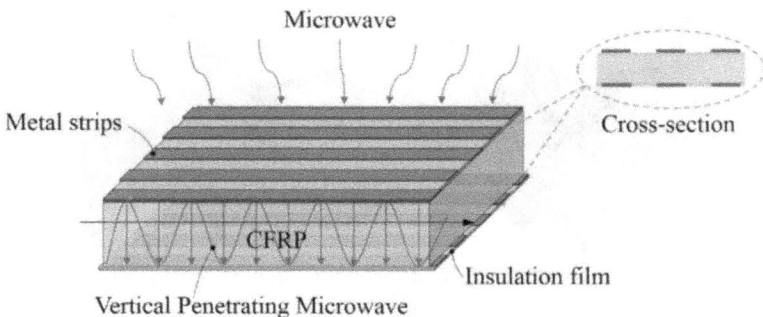

FIGURE 2.10   Vertical penetrating microwave setup (Li et al., 2019).

composites used for the study consisted of bi-directional glass fiber-reinforced in epoxy. It was reported that composites cured at 915 MHz showed maximum flexural strength, that is, 600 MPa, compared to composites cured at 2450 MHz. The interlaminar strength also increased to 9.7% after curing at 915 MHz in comparison to 2450-MHz curing (Rao et al., 2020).

Microwave irradiation techniques have been applied to CNT/polymer composites for the evaluation of the effectiveness of the technique (Arora et al., n.d.). Two types of composites, CNT/PE and CNT/PP, were examined by the uniaxial tensile test, TGA, DSC, XRD, and SEM characterizations. It was observed from the uniaxial tensile test that the Young's modulus of CNT/PE and CNT/PP composites increased to nearly 58% and 47%, respectively (Arora et al., n.d.), compared to virgin polymer. Microwave irradiation led to a decrease in the strength of the composites due to local thermal spots resulting from the high dielectric properties of CNTs and agglomeration.

From SEM examinations, it was revealed that the polymer stretched in the direction of the load along with the bundles of CNTs. The glass transition temperature of the composites has rose to 131 °C and 164 °C as revealed by thermal examination. XRD analysis confirmed the crystallinity of the composites was maintained, which resulted in a good modulus and percentage of elongation. It was also reported that the composites were fabricated within 180–240 s of microwave exposure. The fracture properties, that is, critical stress intensity factor (K_{IC}), of the composites were also evaluated. CNT/PE composites showed an average K_{IC} of 2.08 MPa-m$^{0.5}$ as evaluated using ASTM standard D-5045 (Arora & Pathak, 2020) on a single end notch bend specimen, as shown in Figure 2.11.

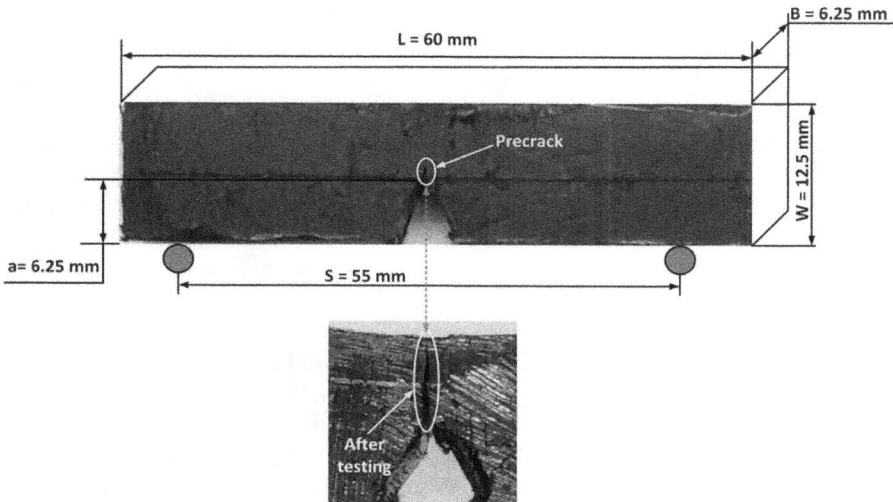

FIGURE 2.11 Single edge notch bend specimen of CNT/PE composite (Arora & Pathak, 2020).

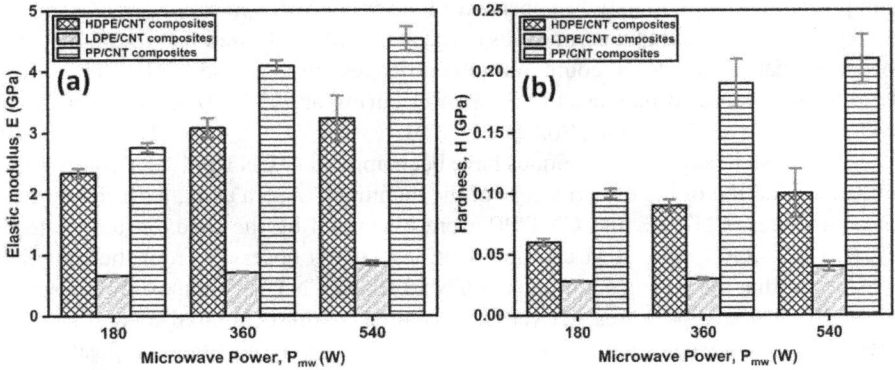

FIGURE 2.12 Variation of (a) elastic modulus and (b) hardness with microwave power (Arora & Pathak, 2021).

The nano-scale properties of microwave-irradiated CNT/polymer composites were evaluated using the nanoindentation technique. CNT/PE and CNT/PP composites were tested using the Hysitron T1970 triboindenter (Arora & Pathak, 2021). The elastic modulus and hardness of three different composites processed using microwave irradiation were evaluated. The hardness and modulus of the composites processed at three different power ratings, 180, 360, and 540 W, are shown in Figure 2.12. CNT/PP composites are nearly 18%, 33%, and 39% stiffer than CNT/PE (high-density polyethylene) composites. However, CNT/PE (high-density polyethylene) composites are nearly 255%, 334%, and 276% stiffer than CNT/PE (low-density polyethylene) composites. The hardness of CNT/PP composites showed an increase from 67% to 110% in comparison to CNT/PE (high-density polyethylene) composites processed from 180 to 54 W microwave power. On the other hand, CNT/PE (high-density polyethylene) composites showed a 114% to 150% increase in comparison to CNT/PE (low-density polyethylene) composites.

The creep response of the composites was also evaluated for microwave-irradiated composites, as shown in Figure 2.13. The viscoelastic properties of the composites were evaluated using the Maxwell and Voigt mathematical models. It was reported that microwave-irradiated CNT/PP composites can store more energy compared to CNT/PE composites. Also, the best fit model for these composites was the Voigt model (Arora & Pathak, 2021).

A comparative study on the nanocomposite film produced by in-situ microwave plasma polymerization was reported (Ibrahim & Was, 2019). The study focused on the use of SWCNTs and MWCNTs in the nanocomposite film in an aniline monomer. The morphology and roughness were characterized by using SEM and AFM. It was reported that these characterization parameters are affected by the type of CNT used. The bonding between PANI and CNTs was confirmed from FTI-IR examination, whereas the homogeneity of the coating of PANI on the CNTs was confirmed by XRD and SEM examination. The conductivity of the nanocomposite film was

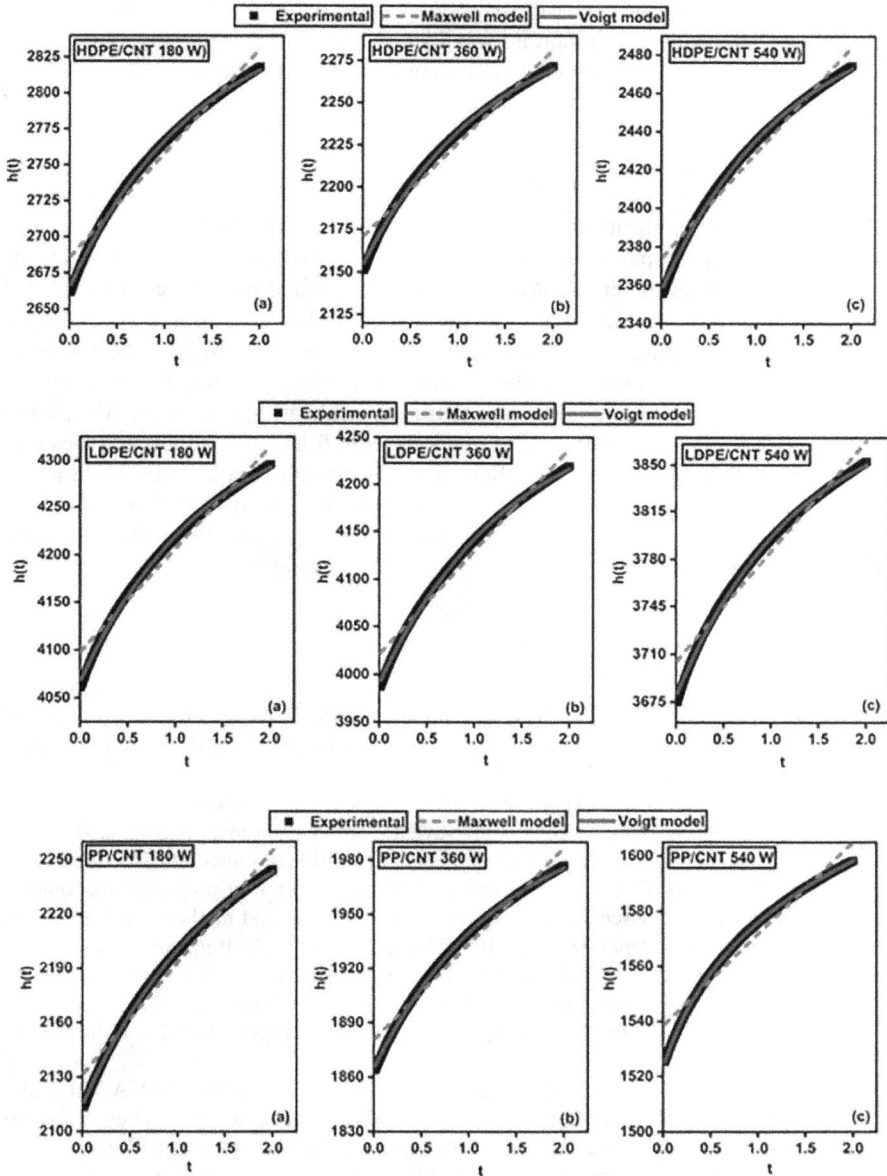

FIGURE 2.13 Creep response of CNT/polymer composites for hold period of 2 s at 10 mN peak load (Arora & Pathak, 2021).

examined by Hall measurement. It was observed that with 1.5% wt. of SWCNTs and MWCNTs, the conductivity recorded was 25 and 7.8 S/cm, respectively, in comparison to virgin PANI, whose conductivity ranged from 8–10 S/cm (Ibrahim & Was, 2019).

2.3 SUMMARY

MIM has gained popularity due to its extraordinary energy and time-saving characteristics. MIM has opened the gate to process polymers with ease and efficiency to replace traditional materials. Also, the MIM technique has proved a successful manufacturing technique for both synthetic and natural filler-reinforced composites. Films and bulk polymer composites have been produced by MIM. The literature on different MIM processes has shown that mechanical, electrical, and magnetic properties are enhanced in comparison to traditional heating processes. Also, it has become an emerging processing technique to recover fillers and then reproduce composites employing these recovered fillers. The composites produced by MIM are in both lamina and laminate form, although some challenges still exist to achieve the enhanced mechanical properties at the expense of others. Owing to these benefits, MIM can be used for the production of PMC at large scale.

REFERENCES

Agrawal, D. (2006). Microwave sintering of ceramics, composites and metallic materials, and melting of glasses. *Transactions of the Indian Ceramic Society*, *65*(3), 129–144. https://doi.org/10.1080/0371750X.2006.11012292

Arora, G., & Pathak, H. (2020). Experimental and numerical approach to study mechanical and fracture properties of high-density polyethylene carbon nanotubes composite. *Materials Today Communications*, *22*, 100829. https://doi.org/10.1016/j.mtcomm.2019.100829

Arora, G., & Pathak, H. (2021). Nanoindentation characterization of polymer nanocomposites for elastic and viscoelastic properties: Experimental and mathematical approach. *Composites Part C: Open Access*, *4*, 100103. https://doi.org/10.1016/j.jcomc.2020.100103

Arora, G., Pathak, H., & Zafar, S. (2019). Fabrication and characterization of microwave cured high-density polyethylene/carbon nanotube and polypropylene/carbon nanotube composites. *Journal of Composite Materials*, *53*(15), 2091–2104. https://doi.org/10.1177/0021998318822705

Arora, G., Singh, M. K., Pathak, H., & Zafar, S. (2021). Micro-scale analysis of HA-PLLA biocomposites: Effect of the interpenetration of voids on mechanical properties. *Materials Today Communications*, *28*, 102568. https://doi.org/10.1016/j.mtcomm.2021.102568

Benítez, R., Fuentes, A., & Lozano, K. (2007). Effects of microwave assisted heating of carbon nanofiber reinforced high density polyethylene. *Journal of Materials Processing Technology*, *190*(1–3), 324–331. https://doi.org/10.1016/j.jmatprotec.2007.02.016

Bhoi, N. K., Singh, H., & Pratap, S. (2019). Microwave material processing: A clean, green, and sustainable approach. In *Sustainable engineering products and manufacturing technologies* (Issue 1). Elsevier Inc. https://doi.org/10.1016/B978-0-12-816564-5.00001-3

Boey, F. Y. C., & Lee, W. L. (1990). Microwave radiation curing of a thermosetting composite. *Journal of Materials Science Letters*, *9*(10), 1172–1173. https://doi.org/10.1007/BF00721880

Dong, S. M., Katoh, Y., Kohyama, A., Schwab, S. T., & Snead, L. L. (2008). Microstructural evolution and mechanical performances of SiC/SiC composites by polymer impregnation/microwave pyrolysis (PIMP) process. *Ceramics International, 28*(2002), 899–905.

Fotiou, I., Baltopoulos, A., Vavouliotis, A., & Kostopoulos, V. (2013). Microwave curing of epoxy polymers reinforced with carbon nanotubes. *Journal of Applied Polymer Science, 129*(5), 2754–2764. https://doi.org/10.1002/app.39003

Gamit, D., Mishra, R. R., & Sharma, A. K. (2017). Joining of mild steel pipes using microwave hybrid heating at 2.45 GHz and joint characterization. *Journal of Manufacturing Processes, 27*, 158–168. https://doi.org/10.1016/J.JMAPRO.2017.04.028

González-Morones, P., Hernández-Hernández, E., Fernández-Tavizón, S., Ledezma-Rodríguez, R., Sáenz-Galindo, A., Cadenas, G., Ávila-Orta, C. A., & Ziolo, R. F. (2018). Exfoliation, reduction, hybridization and polymerization mechanisms in one-step microwave-assist synthesis of nanocomposite nylon-6/graphene. *Polymer, 146*, 73–81. https://doi.org/10.1016/j.polymer.2018.05.014

Gupta, D., Bhovi, P. M., Sharma, A. K., & Dutta, S. (2012). Development and characterization of microwave composite cladding. *Journal of Manufacturing Processes, 14*(3), 243–249. https://doi.org/10.1016/J.JMAPRO.2012.05.007

Gupta, D., & Sharma, A. K. (2014). Microwave cladding: A new approach in surface engineering. *Journal of Manufacturing Processes, 16*(2), 176–182. https://doi.org/10.1016/J.JMAPRO.2014.01.001

Hu, X., Wang, Y., Xu, M., Zhang, L., Zhang, J., & Dong, W. (2018). Mechanical testing and reinforcing mechanisms of a magnetic field-sensitive hydrogel prepared by microwave-assisted polymerization. *Polymer Testing, 71*(August), 344–351. https://doi.org/10.1016/j.polymertesting.2018.09.027

Huang, J., Pen, H., Xu, Z., & Yi, C. (2008). Magnetic Fe_3O_4/poly (styrene-co-acrylamide) composite nanoparticles prepared by microwave-assisted emulsion polymerization. *Reactive and Functional Polymers, 68*, 332–339. https://doi.org/10.1016/j.reactfunctpolym.2007.08.002

Ibrahim, N. I., & Was, A. S. (2019). A comparative study of polyaniline/MWCNT with polyaniline/SWCNT nanocomposite films synthesized by microwave plasma polymerization. *Synthetic Metals, 250*(February), 49–54. https://doi.org/10.1016/j.synthmet.2019.02.007

Jaiswal, K. K., Dhamodaran, M., Ramaswamy, M., & Ramaswamy, A. P. (2017). Microwave-assisted rapid synthesis of Fe_3O_4/Poly(styrene-divinylbenzene-acrylic acid) polymeric magnetic composites and investigation of their structural and magnetic properties. *European Polymer Journal, 98*, 177–190. https://doi.org/10.1016/j.eurpolymj.2017.11.005

Johnsona, M. S., Ruddar, C. D., & Hillb, D. J. (1997). Microwave assisted resin transfer moulding. *Composites Part A Applied Science and Manufacturing, 29*(1–2), 71–86.

Joshi, S. C., & Bhudolia, S. K. (2014). Microwave-thermal technique for energy and time efficient curing of carbon fiber reinforced polymer prepreg composites. *Journal of Composite Materials, 48*(24), 3035–3048. https://doi.org/10.1177/0021998313504606

Kumar Singh, M., Arora, G., Tewari, R., Zafar, S., & Pathak, H. (2022). Materials today: Proceedings effect of pine cone filler particle size and treatment on the performance of recycled thermoplastics reinforced wood composites. *Materials Today: Proceedings, 7853*. https://doi.org/10.1016/j.matpr.2022.02.022

Kumar Singh, M., Verma, N., Kumar, R., Zafar, S., & Pathak, H. (2021). Microwave processing of polymer composites. In *Advanced welding and deforming* (Vol. 347, pp. 351–380). Elsevier. https://doi.org/10.1016/B978-0-12-822049-8.00013-X

Kumar Singh, M., Zafar, S., & Talha, M. (2019). ScienceDirect development of porous biocomposites through microwave curing for bone tissue engineering. *Materials Today: Proceedings, 18*, 731–739. https://doi.org/10.1016/j.matpr.2019.06.478

Lester, E., Kingman, S., Wong, K. H., Rudd, C., Pickering, S., & Hilal, N. (2004). Microwave heating as a means for carbon fibre recovery from polymer composites: A technical feasibility study. *Materials Research Bulletin, 39*(10), 1549–1556. https://doi.org/10.1016/j. materresbull.2004.04.031

Li, S., Louzguine-Luzgin, D. V., Xie, G., Sato, M., & Inoue, A. (2010). Development of novel metallic glass/polymer composite materials by microwave heating in a separated H-field. *Materials Letters, 64*(3), 235–238. https://doi.org/10.1016/j.matlet.2009.10.017

Li, Y., Li, N., Zhou, J., & Cheng, Q. (2019). Microwave curing of multidirectional carbon fiber reinforced polymer composites. *Composite Structures, 212*(November 2018), 83–93. https://doi.org/10.1016/j.compstruct.2019.01.027

Lin, W., Moon, K. S., & Wong, C. P. (2009). A combined process of in situ functionalization and microwave treatment to achieve ultrasmall thermal expansion of aligned carbon nanotube-polymer nanocomposites: Toward applications as thermal interface materials. *Advanced Materials, 21*(23), 2421–2424. https://doi.org/10.1002/adma.200803548

Mishra, R. R., & Sharma, A. K. (2016). Microwave-material interaction phenomena: Heating mechanisms, challenges and opportunities in material processing. *Composites Part A: Applied Science and Manufacturing, 81*, 78–97. https://doi.org/10.1016/j.compositesa.2015.10.035

Moulart, A., Marrett, C., & Colton, J. (2004). Polymeric composites for use in electronic and microwave devices. *Polymer Engineering & Science, 44*(3), 588–597. https://doi.org/10.1002/PEN.20053

Nahas, M. N., Jilani, A., & Salah, N. (2016). Microwave synthesis of ultrathin, non-agglomerated CuO nanosheets and their evaluation as nanofillers for polymer nanocomposites. *Journal of Alloys and Compounds, 680*, 350–358. https://doi.org/10.1016/j. jallcom.2016.04.147

Oh, W., Nahar, K., Youn, K., Ud, R., & Biswas, D. (2020). Microwave-assisted synthesis of conducting polymer matrix based thin film NaLa (MoO 4) 2-G-PPy composites for high-performance gas sensing. *Surfaces and Interfaces, 21*(September), 100713. https://doi.org/10.1016/j.surfin.2020.100713

Padmanabhan, S., Gupta, A., Arora, G., Pathak, H., Burela, R. G., & Bhatnagar, A. S. (2020). Meso—macro-scale computational analysis of boron nitride nanotube-reinforced aluminium and epoxy nanocomposites: A case study on crack propagation. *Proceedings of the Institution of Mechanical Engineers, Part L: Journal of Materials: Design and Applications, 235*(2), 1–16. https://doi.org/10.1177/1464420720961426

Poyraz, S., Zhang, L., Schroder, A., & Zhang, X. (2015). Ultrafast microwave welding/reinforcing approach at the interface of thermoplastic materials. *CS Applied Materials & Interfaces, 7*(40). https://doi.org/10.1021/ACSAMI.5B06484

Pramila Devi, D. S., Nair, A. B., Jabin, T., & Kutty, S. K. N. (2012). Mechanical, thermal, and microwave properties of conducting composites of polypyrrole/polypyrrole-coated short nylon fibers with acrylonitrile butadiene rubber. *Journal of Applied Polymer Science, 126*(6), 1965–1976. https://doi.org/10.1002/APP.36924

Pundhir, N., Zafar, S., & Pathak, H. (2019). Performance evaluation of HDPE/MWCNT and HDPE/kenaf composites. *Journal of Thermoplastic Composite Materials, 34*(10). https://doi.org/10.1177/0892705719868278

Rao, S., Vijapur, L., & Prakash, M. R. (2020). Effect of incident microwave frequency on curing process of polymer matrix composites. *Journal of Manufacturing Processes, 55*(July 2019), 198–207. https://doi.org/10.1016/j.jmapro.2020.04.003

Singh, M. K., Verma, N., Pundhir, N., Zafar, S., & Pathak, H. (2020). Optimization of microwave power and reinforcement in microwave-cured coir/HDPE composites. In Biswal, B., Sarkar, B., & Mahanta, P. (eds.) *Advances in mechanical engineering. lecture*

notes in mechanical engineering (pp. 159–171). Springer, Taylor & Francis. https://doi.
  org/10.1007/978-981-15-0124-1_16

Singh, M. K., Verma, N., & Zafar, S. (2019). Optimization of process parameters of microwave
  processed PLLA/coir composites for enhanced mechanical behavior. *Journal of Physics:
  Conference Series, 1240*, 012038. https://doi.org/10.1088/1742-6596/1240/1/012038

Singh, M. K., Verma, N., & Zafar, S. (2020). Conventional processing of polymer matrix composites. In *Lightweight polymer composite structures: Design and manufacturing techniques* (pp. 21–66). CRC Press, Taylor & Francis. https://doi.org/10.1201/9780429244087;
  https://www.taylorfrancis.com/chapters/edit/10.1201/9780429244087-2/conventional-
  processing-polymer-matrix-composites-singh-verma-zafar

Singh, M. K., & Zafar, S. (2018). Influence of microwave power on mechanical properties
  of microwave-cured polyethylene/coir composites. *Journal of Natural Fibers, 17*(6),
  845–860. https://doi.org/10.1080/15440478.2018.1534192

Singh, M. K., & Zafar, S. (2019). Development and mechanical characterization of microwave-
  cured thermoplastic based natural fibre reinforced composites. *Journal of Thermoplastic
  Composite Materials, 32*(10), 1427–1442. https://doi.org/10.1177/0892705718799832

Singh, M. K., & Zafar, S. (2020a). Abrasive wear mechanism of microwave-assisted compression molded kenaf/HDPE composite. *Journal of Tribology, 142*(10), 1–11. https://doi.
  org/10.1115/1.4046858

Singh, M. K., & Zafar, S. (2020b). Effect of layering sequence on mechanical properties of
  woven kenaf/jute fabric hybrid laminated microwave-processed composites. *Journal of
  Industrial Textiles*, 1–22. https://doi.org/10.1177/1528083720911219

Singh, M. K., & Zafar, S. (2021). Wettability, absorption and degradation behavior of microwave-assisted compression molded kenaf/HDPE composite tank under various environments. *Polymer Degradation and Stability, 185*, 109500.

Singh, M. K., Zafar, S., & Talha, M. (2019). Development and characterisation of poly-L-
  lactide-based foams fabricated through microwave-assisted compression moulding.
  Journal of Cellular Plastics, 55(5), 523–541. https://doi.org/10.1177/0021955X19850728

Singh, S., Gupta, D., & Jain, V. (2018). Processing of Ni-WC-8Co MMC casting through
  microwave melting. *Materials and Manufacturing Processes, 33*(1), 26–34. https://doi.
  org/10.1080/10426914.2017.1291954

Singla, P., Mehta, R., & Upadhyay, S. N. (2014). Microwave assisted in situ ring-opening
  polymerization of polylactide/clay nanocomposites: Effect of clay loading. *Applied
  Clay Science, 95*, 67–73. https://doi.org/10.1016/j.clay.2014.03.012

Skorokhod, V. V. (1995). Theory of the physical properties of porous and composite materials
  and the principles for control of their microstructure in manufacturing processes. *Powder
  Metallurgy and Metal Ceramics, 34*(1), 48–63. https://doi.org/10.1007/BF00559852

Spasojevi, P. (2013). Unique effects of microwave heating on polymerization kinetics of poly
  (methyl methacrylate) composites. *Materials Chemistry and Physics, 141*, 882–890.
  https://doi.org/10.1016/j.matchemphys.2013.06.019

Tamang, S., & Aravindan, S. (2017). An investigation on joining of Al6061-T6 to AZ31B
  by microwave hybrid heating using active braze alloy as an interlayer. *Journal of
  Manufacturing Processes, 28*, 94–100. https://doi.org/10.1016/J.JMAPRO.2017.05.027

Taylor, P., Singh, S., Gupta, D., Jain, V., & Sharma, A. K. (2015). Microwave processing
  of materials and applications in manufacturing industries: A review. *Materials and
  Manufacturing Processes, 30*(1), 37–41. https://doi.org/10.1080/10426914.2014.952028

Tewari, R., Singh, M. K., & Zafar, S. (2021). Utilization of forest and plastic wastes for composite manufacturing using microwave-assisted compression molding for low load applications. *Journal of Polymer Research, 28*(11). https://doi.org/10.1007/s10965-021-02778-6

Tewari, R., Singh, M. K., Zafar, S., & Powar, S. (2020). Parametric optimization of laser drilling of microwave-processed kenaf/HDPE composite. *Polymers and Polymer Composites, 29*(3), 176–187. https://doi.org/10.1177/0967391120905705

Thostenson, E. T., & Chou, T. (1999). Microwave processing: Fundamentals and applications. *Composites Part A: Applied Science and Manufacturing, 30*(9), 1055–1071.

Wang, C., Chen, T., Chang, S., Cheng, S., & Chin, T. (2007). Strong carbon-nanotube-polymer bonding by microwave irradiation. *Advanced Functional Materials, 17*(12), 1979–1983. https://doi.org/10.1002/adfm.200601011

Wise, R. J., & Froment, I. D. (2001). Microwave welding of thermoplastics. *Journal of Materials Science, 36*(24), 5935–5954. https://doi.org/10.1023/A:1012993113748

Xie, R., Wang, J., Yang, Y., Jiang, K., Li, Q., & Fan, S. (2011). Aligned carbon nanotube coating on polyethylene surface formed by microwave radiation. *Composites Science and Technology, 72*(1), 85–90. https://doi.org/10.1016/j.compscitech.2011.10.003

Zabihi, O., Ahmadi, M., Liu, C., Mahmoodi, R., Li, Q., & Naebe, M. (2020). Development of a low cost and green microwave assisted approach towards the circular carbon fibre composites. *Composites Part B, 184*(November 2019), 107750. https://doi.org/10.1016/j.compositesb.2020.107750

Zhang, J., Duan, Y., Wang, B., & Zhang, X. (2020). Interfacial enhancement for carbon fibre reinforced electron beam cured polymer composite by microwave irradiation. *Polymer*, 122327. https://doi.org/10.1016/j.polymer.2020.122327

3 Microwave Joining
A Sustainable Manufacturing Technique

Saloni, Sarbjeet Kaushal

CONTENTS

3.1 INTRODUCTION

Joining has applications in every field, be it manufacturing of huge aircraft to small microprocessors (Agrawal, n.d.). Metallic pipes are manufactured using a variety of casting or forming processes. Further, to accommodate the complex flow passages in various industrial applications, a wide variety of joining techniques are used (Silberalitt et al., 1993). To join metallic pipes in industries, welding technologies such as tungsten inert gas (TIG) welding, gas metal arc welding (GMAW), friction welding, and laser beam welding (LBW) are often employed (Silberglitt, 1995). The joints produced through these joining methods exhibit excellent strength and satisfactory service life (Cai et al., 1992). However, in recent years, global industry standards are demanding energy-efficient, time-saving, and ecofriendly joining techniques that can cater to the needs of sustainable manufacturing industry (Chandrasekaran et al., 2011).

As the need of the hour, many researchers have started exploring methods of sustainable material processing and manufacturing (Badiger et al., 2015). Microwave

DOI: 10.1201/9781003269298-3

33

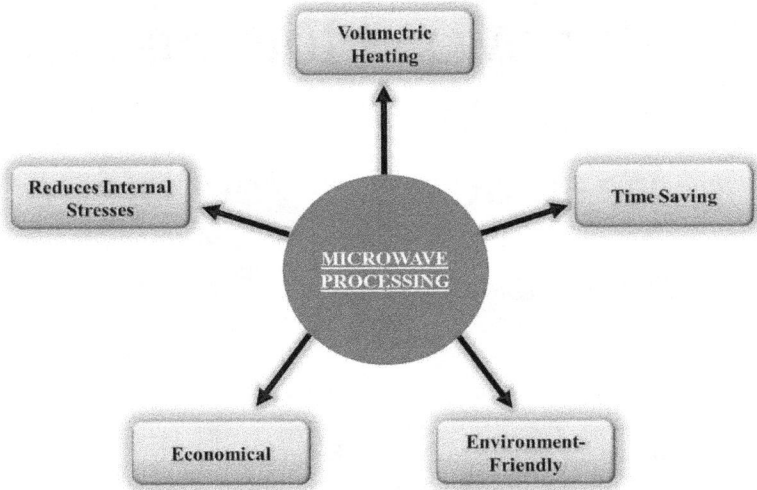

FIGURE 3.1   Advantages of microwave processing (Kamble et al., 2021; Sahota et al., 2020).

material processing has emerged as a sustainable method of material processing in recent years. Microwave processing is growing rapidly in cladding, sintering, joining, casting, and other applications (Silberalitt et al., 1993). Microwave material processing has advantages over conventional methods of material processing such as its environmentally friendly nature, lower processing time, and economy (Bansal et al., 2015). It provides volumetric heating to the material, which heats the material uniformly, giving rise to uniform grain size and fewer heat-affected zones and relieves internal stresses. Figure 3.1 shows the various advantages of microwave heating. However, it is difficult to process metals through microwave radiation at ambient temperature, as metals reflect back radiation due to their lower penetration depth at room temperature, which restricts metals in absorbing microwaves (Cozzi et al., 2008). However, researchers have worked on methods to increase metal penetration depth using susceptor materials (like charcoal powder), which help to heat the metal conventionally first to increase its penetration depth. With an increase in penetration depth, the metal particles start absorbing microwave radiation (Kumar & Sehgal, 2021b).

3.2   METHODS OF JOINING THROUGH MICROWAVE RADIATION

Bulk metals are joined with microwave irradiation through numerous techniques. As metals reflect microwaves, it becomes difficult to join them (Gautam & Vipin, 2021). So researchers have worked on various alternative approaches to produce joints using microwave radiation (Singh et al., 2014). To date, researchers have successfully joined metals using three techniques with microwave radiation, which are discussed in the following.

3.2.1 Direct Microwave Heating

Siores and Do Rego (1995) initially worked on joining metals through microwave radiation. In the experiment, the metallic pieces were directly introduced to microwave radiation. The authors made a butt joint of thin metal steel sheets of thickness 0.1–0.3 mm. A small space of ~0.2 mm was maintained in the sheets. The metal sheets were exposed to microwaves of power 2 kW. As the metal sheets were directly exposed to microwaves, they reflected back the radiation in the form of arcs and flashes, which were produced at the interface gap. These flashes and arcs helped in localized melting of sheets. The light pressure helped in fusion of the material. The experimental setup is shown in Figure 3.2 (Siores & Do Rego, 1995).

3.2.2 Microwave Hybrid Heating

Joining through direct microwave heating is effective for very thin sheets. Many researchers have started working on methods for joining metals using microwave energy without reflecting radiation back and getting uniform volumetric heating (Srinath et al., 2018). This was achieved by a microwave hybrid heating (MHH) technique. This method uses susceptor powder to absorb the radiation and avoids direct contact of radiation with the metal (Badiger et al., 2018b). This technique involves conventional and non-conventional heating. Initially the skin depth of the metal is small, which prevents the metals absorbing radiation; therefore, first, the susceptor absorbs the radiation and heats up the interface of the joint through conventional heating. This conventional heating in turn increases the skin depth of the metal, and the metal starts absorbing microwaves, which leads to volumetric uniform heating of the joint, which produces a joint with less internal stress and uniform grain size. The heat affected zone (HAZ) is smaller. Figure 3.3 depicts a schematic diagram of MHH.

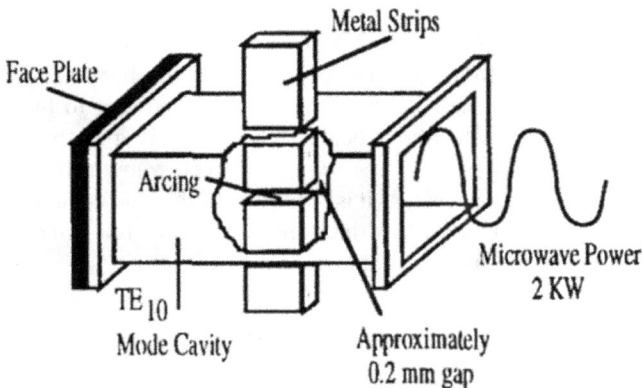

FIGURE 3.2   Microwave welding of metals by arcing (Siores & Do Rego, 1995).

Source: (Reproduced with the permission of rightsholder)

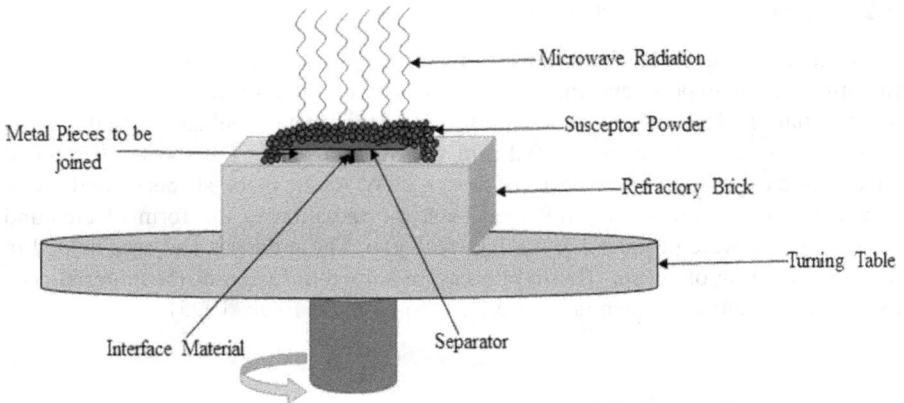

FIGURE 3.3   Schematic representation of joining through microwave hybrid heating (Kamble et al., 2021; Nandwani et al., 2019).

3.2.2.1 Role of Susceptor Powder in MHH

In MHH, susceptor powder plays an important role. It helps in conventional heating of the metal workpiece so as to increase the skin depth of the metals. It raises the temperature of the workpiece. With an increase in skin depth, the material starts absorbing microwave irradiation. Most research is carried out with charcoal as a susceptor powder. Charcoal is easily available and economical, which provides better heating of material. The thickness of the charcoal layer should be optimized, as with a thin layer, it will reduce the rate of conventional heating of the material, and a thick layer of charcoal powder will reduce the amount of microwave radiation reaching the metal pieces, which results in more conventional heating than microwave heating. The thickness of the charcoal layer will affect the exposure time of the joint.

3.2.2.2 Role of Separator Sheet in MHH

A separator sheet used in MHH avoids intermixing of the charcoal powder with the filler material. It transfers the heat from the susceptor powder to the metal pieces (Mishra & Sharma, 2016). The rate of heat transfer depends upon the type of separator sheet used and the thickness of the separator sheet. Many researchers work by using graphite as a separator sheet, as it is readily available and economical and helps in transfer of the heat to the metal. A thick sheet decreases the rate of heat transfer, hence increasing the exposure time (Samyal et al., 2019b).

3.2.3 Selective Microwave Hybrid Heating

Researchers have worked on improving methods for joining metals using microwave radiation. Joining through the MHH method was further improved by selective microwave hybrid heating (SMHH). In this method, the specimens (or workpieces) to be joined are kept on a refractory brick, and the interface material is added at the

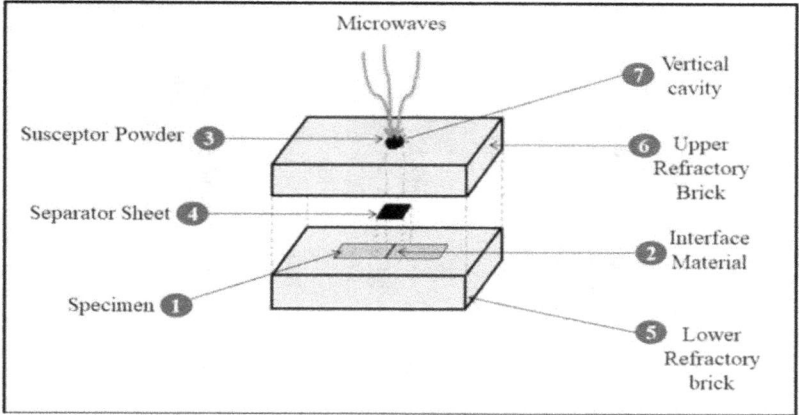

FIGURE 3.4   Visual presentation of selective microwave hybrid heating (Naik et al., 2021).

Source: (Reproduced with the permission of rightsholder)

joint region. The joint is covered by the separator (graphite sheet). The workpieces are further covered by refractory brick with a vertical cavity to the joint area. The susceptor is poured into the cavity to increase the joining ability. This process helps to decrease the heat-affected zone. It targets the joint area only, which decreases the exposure time for producing a microwave joint. A diagrammatic representation of SMHH is shown in Figure 3.4.

3.3   INTERFACIAL MATERIAL USED

In microwave joining of materials, interfacial materials play an important role, and various kinds of interfacial materials have been used in microwave joining process. For joining of metal pieces, it is required to determine the type of interfacial material used to make the joint. Researchers have worked on optimizing the properties of the joint by using different types of interfacial materials for joining different/similar metals. Also, the particle size of the interfacial material used has an effect on exposure time of the joint to microwave irradiation.

The effects of different interfacial material are provided in Table 3.1.

Many researchers have developed joints without the use of any interfacial metal, which reduces the cost of producing the joint. V. Kumar and Sehgal (2020) developed an economically effective joint of duplex stainless steel (SS2205) without using any filler material. It was observed that the microhardness of the joint, 410 HV, was more than that of the base metal because of the formation of carbides of nickel, iron, and chromium. The formation of carbides was due to the reaction with carbon from a graphite sheet, which was used to separate the joint from susceptor powder (V. Kumar & Sehgal, 2020). A. Kumar et al. (2019) joined stainless steel SS304 with stainless steel SS316 using a MHH technique without any filler powder. It was found that the microhardness of the joint formed was 289.8 HV, which was more than that

TABLE 3.1

Effect of Interface Materials on the Properties of Microwave-Processed Joints (continued)

Sr. No.	Interface Material	Base Material	Microhardness Comparison of Joint with Base Metal	Conclusion
1.	Epoxy resin and copper powder slurry (Srinath et al., 2011a)	Copper-copper (Srinath et al., 2011a)	Microhardness of joint was 84% of bulk copper (Srinath et al., 2011a)	About 26% of copper powder at the joint was converted into copper oxide, which improved coupling of microwaves with powder. Microstructure of the joint is denser, with 1.96% porosity (Srinath et al., 2011a).
2.	Epoxy resin and pure nickel slurry (Tamang & Aravindan, 2017)	Aluminum (Al) alloy Al6061-T6–magnesium (Mg) alloy AZ31B (Tamang & Aravindan, 2017)	Hardness of joint was 4.31 GPa, that of base alloy of Al was 0.64 GPa, and Mg alloy was 0.60 GPa (Tamang & Aravindan, 2017)	The presence of oxides and intermetallic compounds resulted in low strength (Tamang & Aravindan, 2017).
3.	Epoxy resin and EWAC powder slurry (Satnam Singh et al., 2019)	Cast iron-cast iron (Satnam Singh et al., 2019)	Vickers hardness of joint (201.7 ± 18 HV) on contrary with base metal (185 ± 4 HV) (Satnam Singh et al., 2019)	The porosity of the joint was 1.5%–1.88%. The joints formed have tensile strength of 90% of the bulk material. Formation of high-strength intermetallic compounds leads to high strength (Satnam Singh et al., 2019).
4.	No interfacial material was used (V. Kumar & Sehgal, 2020)	Duplex stainless steel SS2205-duplex stainless steel SS2205 (V. Kumar & Sehgal, 2020)	Microhardness of joint was 410 HV, and microhardness of base metal was 252 HV (V. Kumar & Sehgal, 2020)	Hardness of joint region was more due to absorption of charcoal from susceptor placed above the joint (V. Kumar & Sehgal, 2020).
5.	Stainless steel SS316 (S. Kumar et al., 2020)	Stainless steel SS304-stainless steel SS316 (S. Kumar et al., 2020)	Microhardness of joint was 380 HV, and microhardness of base metal was 200 ± 20 HV (S. Kumar et al., 2020)	Joint formed was homogeneous and hardness of joint was more due to formation of carbides of chromium, nickel, and iron. Crack-free joint was formed (S. Kumar et al., 2020).

TABLE 3.1

Effect of Interface Materials on the Properties of Microwave-Processed Joints (continued)

	Interface material	Joint	Microhardness	Remarks
6.	Nickel powder and blumer 1400XX slurry (Soni et al., 2018)	Stainless steel SS316-stainless steel SS316 (Soni et al., 2018)	Microhardness of joint was 450 HV as compared to base material with 152 HV (Soni et al., 2018)	Tensile strength of nano-powder–based joints is 24% higher than their micro size powder. Use of nano-powder helps in increasing hardness of the joints by 7% (Soni et al., 2018).
7.	No interfacial material was used (Bagha et al., 2019)	Stainless steel SS304-stainless steel SS304 (Bagha et al., 2019)	Microhardness of joint was 260 HV, and that of base metal was 198 HV (Bagha et al., 2019)	Ultimate tensile strength of the joints was 271 MPa (Bagha et al., 2019).
8.	Nickel powder and epoxy resin slurry (Gamit et al., 2017)	Mild steel pipe-mild steel pipe (Gamit et al., 2017)	Microhardness of the joint was 572 HV, and that of base metal was 397 HV (Gamit et al., 2017)	Porosity at the outer surface of the joint was higher as compared to the inner surface due to the presence of gases due to heating of susceptor (Gamit et al., 2017).
9.	EWAC powder and epoxy resin slurry (Satnam Singh et al., 2017)	Hastelloy-hastelloy (Satnam Singh et al., 2017)	Microhardness of the joint was 708 ± 10 HV, whereas that of base metal was 655 ± 5 HV (Satnam Singh et al., 2017)	Microwave produced joints of hastelloy showed 82.13% strength of base metal (524 MPa) (Satnam Singh et al., 2017).
10.	No interfacial material was used (A. Kumar et al., 2019)	Stainless steel SS304-stainless steel SS316 (A. Kumar et al., 2019)	Microhardness of the joint was 289.8 HV, that of SS304 was 207 HV, and SS316 was 182 HV (A. Kumar et al., 2019)	Joints were crack free and had homogenous microstructures due to volumetric heating (A. Kumar et al., 2019).

of the base metal (A. Kumar et al., 2019). Bagha et al. (2019) successfully prepared a joint of stainless steel (SS304-SS304) without using any interface material, for the purpose of making it cost effective. It was observed that the joint formed without using interface powder showed better properties compared to those produced with the use of interface material of particle size 50–40 μm. On the other hand, the joint formed using powder of particle size 30—20 μm had higher tensile properties as compared to a powder-free joint. It was observed that in reducing the cost of making the joint, the tensile strength was lower as compared to joints formed using interfacial material with smaller powder size (Bagha et al., 2019).

Most researchers have worked on developing a joint using a slurry of nickel-based powder with epoxy resin as an interface material. Epoxy resin helps provide adhesion to the joint. At high temperatures, the nickel-based powder melts and epoxy resin evaporates, thus forming the joint (P. Singh et al., 2021). Srinath et al. (2011a, 2011b) prepared a joint of stainless steel (SS316) with mild steel using a nickel-based powder with particle size of 40 μm. It was found that the microhardness of the joint was 133 Hv, and the ultimate tensile strength of the joint was 346.6 MPa. It was found that there was formation of metallic carbides ($Cr_{23}C_6$) and cementite (Fe_3C) at the joint region (Srinath et al., 2011b). Badiger et al. (2015) worked out joining an Inconel 625 alloy using the MHH technique using nickel as interface powder. Formation of chromium carbides increased the microhardness of the joint surface, which was observed as 350 ± 10 HV. The carbide formation also affected the tensile strength of the joint surface, which was ~35% of the base metal (Badiger et al., 2015). Gamit et al. (2017) successfully formed a joint of mild steel pipes using nickel-based powder with particle size of 40 μm as interface material. It was found that the microhardness of the joint was 572 HV, and the microhardness of the pipe was 397 HV. The reason for the higher hardness at the joint was due to the formation of carbides and intermetallic Ni (Gamit et al., 2017). Badiger et al. (2018a) developed a joint of Inconel 625 through MHH with the use of nickel-based EWAC powder with a particle size of 53 and 75 μm, respectively. It was found that the ultimate tensile strength was the maximum with a particle size of 53 μm, at 345 MPa (Badiger et al., 2018a). Singh et al. (2019) successfully joined cast iron plates through microwave radiation using a slurry of nickel-based EWAC powder and epoxy resin as interface material. It was found that the tensile strength of the joint formed was nearly 90% that of the base metal (cast iron). This high strength is the result of the formation of intermetallic and the use of nickel, which is ductile in nature as an interfacial material. The porosity of the joint was lower than that of the base metal because of uniform volumetric heating of the joint with microwave radiation (Satnam Singh et al., 2019).

Soni et al. (2018) successfully conducted an experiment for joining stainless steel (SS316) using MHH with nickel nano-powder as interface material and compared the results with those of using nickel micro-powder. It was found that the joint formed with nano-powder was harder (450 HV) as compared to that formed using micro-sized nickel powder (420 HV). Also, the ultimate tensile strength of the joint formed using nano-powder (385 MPa) was more than that of the joint formed by micro-sized powder (309 MPa), and the joint formed by nano-powder was homogenous and crack free. Thus, nano-sized nickel particles used as interface material are found to be more efficient than micro-sized nickel particles as interface material (Soni et al., 2018). Tamang et al. (2019)

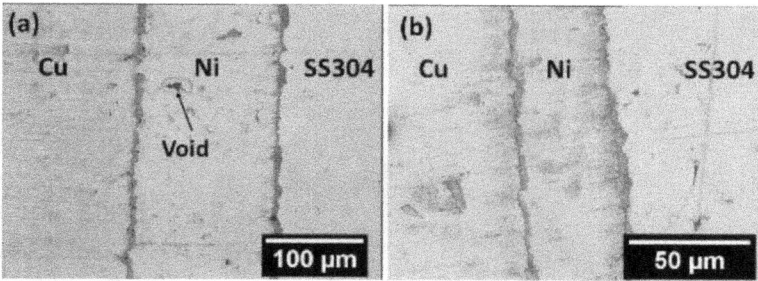

FIGURE 3.5   Optical micrograph of the microwave-joined Cu to SS034 samples (a) 45 μm Ni; (b) 200 nm Ni powder (Tamang & Aravindan, 2019).

Source: (Reproduced with the permission of rightsholder)

(Tamang & Aravindan, 2019) successfully conducted an experiment on joining copper with stainless steel (SS304) through MHH with nickel as the interface material. The results of using 45-μm and 200-nm Ni powder were studied. Figure 3.5 shows optical micrographs of the joint formed by the 200-nm and 40-μm powder sizes. The joint formed by 200-nm nickel powder was formed faster as compared to 45-μm interface powder. Also, the microhardness and strength of the 200-nm-sized powder were greater than those of the 45-μm-sized interface powder (Tamang & Aravindan, 2019).

Bagha et al. (2017) developed a joint of SS304 using the selective hybrid carbon microwave joining (SHCMJ) technique using a slurry of nickel powder and Blumer. A comparison of joints formed using different sizes of nickel powder (50, 40, 30, and 20 μm) was done. Scanning electron microscopy (SEM) images of joints formed with different materials are shown in Figure 3.6. It was observed that with a decrease in particle size, the ultimate tensile strength increased, but on the contrary, the ductility of the joint decreased. Also, with smaller particle size, the homogeneity of the joint also increased (Bagha et al., 2017).

Gupta, Chaudhary, et al. (2021) fabricated a brass joint through a MHH technique using a slurry of epoxy resin with nickel powder of particle size 40 μm as interface material. A strong joint was formed with an exposure time of 480 seconds. It was found that the joint had an ultimate tensile strength of 292.67 MPa and microhardness of 50 HV (Gupta, Chaudhary, et al., 2021). Gupta, Chaudhary et al. (2021) (Gupta, Kumar, et al., 2021) worked on joining mild steel (MS1020) plates using microwave radiation with nickel as interfacial material. The thickness of the interface was kept as 0.2 mm, with an exposure time of 700 seconds to the microwaves. The porosity, ultimate tensile strength, and hardness of the joint were found to be 1.86%, 247.2 MPa, and 180 Hv, respectively (Gupta, Kumar, et al., 2021). Samyal et al. (2021) prepared a lap joint of stainless steel SS202 using nickel-based powder with a particle size of 20 μm used as filler metal. It was found that the Vickers hardness of the joint formed was ~200 HV, which was approximately equal to that of the base metal (Samyal et al., 2021). Pal et al. (2020) experimented on joining stainless steel SS304 with stainless steel SS316 with MHH using a nickel-based powder as

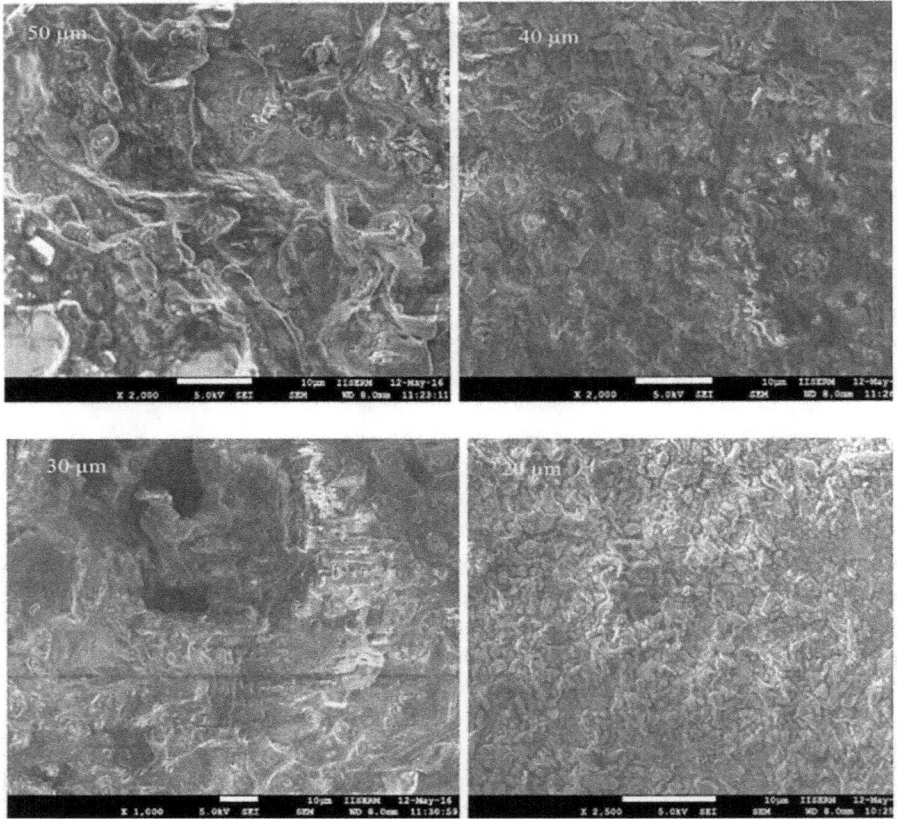

FIGURE 3.6   SEM images of 50, 40, 30, and 20 μm powder-based weld bead (Bagha et al., 2017).

Source: (Reproduced with the permission of rightsholder)

interface material. The effect of the joint formed was studied using different particle sizes of interface material: 70 nm, 20 μm, and 50 μm. It was observed that the maximum carbon content was found in the joint formed with the 70-nm particle size, whereas the joint formed with a 50-μm particle size contained the lowest concentration of carbon content, affecting the hardness of the joint formed (Pal et al., 2020).

Many researchers have also experimented with joining metals using interface powders of the same base metal, which might exhibit similar characteristics to the base metal. Srinath et al. (2011a, 2011b) developed a joint of bulk copper metal using microwave radiation. Using a slurry of copper powder and epoxy resin as interface material, a joint with no cracks was formed, indicating better joint efficiency. Coupling of microwaves was improved by the formation of copper oxides with the heat incoming from the microwaves, which also changed the atomic structure of the copper. The microhardness of the

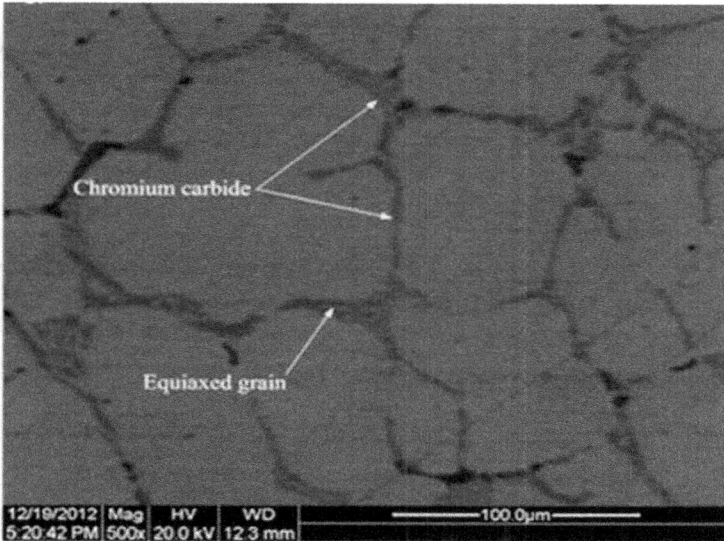

FIGURE 3.7 Presence of carbides of chromium at grain boundaries (Bansal et al., 2014).

Source: (Reproduced with the permission of rightsholder)

joint was ~84% that of the bulk copper metal. Thus, the joint formed showed high tensile strength due to good bonding between the interface surfaces (Srinath et al., 2011a). Bansal et al. (2014) successfully prepared a butt joint of stainless-steel plates (SS-316) using microwave irradiation using stainless steel powder (~50 μm) as interfacial material. It was observed that chromium carbides were formed at the grain boundaries; these carbides tend to increase the hardness of the joint. The ultimate tensile strength of the joint was found to be 82.5% of the base metal (Bansal et al., 2014).

Singh et al. (2015) formed lap and butt joint of technically difficult-to-join metal, aluminum, with the help of microwave irradiation using aluminum powder with particle size ~45 μm used as interface material. It was found that the joint formed with an exposure time of 10 minutes had a hardness of 72.4 ± 10 HV (Shivinder Singh et al., 2015).

3.4 JOINING OF DISSIMILAR METALS

Srinath et al. (2011a, 2011b) performed the joining of stainless steel (SS-316) to mild steel using a multimode applicator oven at a frequency of 2.45 GHz at 900 W. The authors studied various mechanical and metallurgical properties of the joint formed using the MHH technique. A slurry of nickel-based powder of particle size 40 μm with epoxy resin (Bisphenol-A, Blumer 1450XX) was used as an interfacial material, and to prevent bulk metal from microwave radiation, charcoal powder was used as susceptor. It was exposed to microwave radiation for 450 s (Srinath et al., 2011b). Tamang and Aravindan (2017) successfully prepared a joint of aluminum alloy (Al6061-T6) with magnesium alloy (AZ31B) through the MHH method in a domestic oven at

700 W power. An active Braze alloy (TiCuSil) paste was used as interface material for joining the alloy. The authors experimented with using different cavities like alumina-silica, alumina, and graphite and different susceptors like charcoal, SiC, and graphite powders. It was observed that the best results were obtained from graphite cavity and graphite powder as susceptor powder. An exposure time of 525 seconds produced a better joint. It was found that the hardness of the joint (4.31 MPa) was higher than that of the base metal (Tamang & Aravindan, 2017). Tamang and Aravindan (2019) prepared a joint of dissimilar metals: Cu with stainless steel (SS304) in microwave oven at 1280 W using a slurry of epoxy resin and nickel powder. The pieces were kept in a graphite crucible, and graphite susceptor powder was used to initiate microwave heating of the joint. X-ray diffraction (XRD) analysis confirmed the presence of $FeNi_3$ at the joint region (Tamang & Aravindan, 2019). S. Kumar et al. (2020), Pal et al. (2020), and A. Kumar et al. (2019) performed experimentation on joining different grades of stainless steels, that is, SS304 with SS316. S. Kumar et al. (2020) used SS316 powder as an interfacial material, Pal et al. (2020) used nickel powder as an interface powder, and A. Kumar et al. (2019) did not use any filler metal at the interface for making the joint. It was found that with SS316 interface powder, an exposure time of 360 s produced a good joint with microhardness of 420 HV, whereas the joint formed without the use of any interfacial powder had a microhardness of 289.8 HV with an exposure time of 460 s (A. Kumar et al., 2019; S. Kumar et al., 2020; Pal et al., 2020).

3.5 EFFECT OF EXPOSURE TIME ON MICROWAVE JOINING

Exposure time affects the properties of the joint formed. The exposure time should be optimized so as to make a proper joint without any defects. Researchers worked on different trials for performing experimentation with different exposure times.

Gupta, Kumar, et al. (2021) performed experiments on joining mild steel pieces using microwave energy. The authors studied the effect of exposure time in the range of 100–900 s with an interval of 100 s. It was found that up to 300 s of exposure time, no joint was formed, but with an increase in exposure time to 400–500 s, there was improper fusion of the joint. With exposure time of 700 s, a strong joint was formed. But with the increase in time, the joint started distorting, and ultimately melting took place. With an optimal time of 700 s, a joint with tensile strength 495 MPa and microhardness 250 HV was formed (Gupta, Kumar, et al., 2021). Gamit, Mishra, et al. (2017) carried out experimentation on joining mild steel pipes through microwave irradiation. With continuous trials with exposure time of 420, 480, and 510 s it was found that the efficiency of the joint was 40%, 65%, and 51%, respectively. Hence, the optimal exposure time for joining pipes was 480 s (Gamit, Mishra, et al., 2017).

Samyal et al. (2019a, 2019b) successfully conducted an experiment on joining different grades of stainless steel (SS202-SS202, SS304-SS304, and SS316-SS316) using microwave selective hybrid microwave joining with 99.9% pure nickel powder (20 µm particle size) as interface material and studied the exposure time for all these grades. It was found that at a nickel-Blumer ratio of 75:25, the thickness of the interface should be 0.5 mm, and the exposure time for SS202-SS202 was 270 seconds, SS304-SS304 was 370 seconds, and SS316-SS316 was 330 seconds (Samyal et al., 2019a). Table 3.2 shows the exposure times for different types of joints.

TABLE 3.2

Exposure Time Studies for Different Joints

Sr. No	Power and Frequency	Size of Base Metal	Interface Material	Base Metal	Exposure Time
1.	900 W and 2.45 GHz (Srinath et al., 2011a)	15 × 12 ×4 mm and ø18 × 12 mm (Srinath et al., 2011a)	Epoxy resin and copper powder slurry (Srinath et al., 2011a)	Copper-copper (Srinath et al., 2011a)	900 s (Srinath et al., 2011a)
2.	700 W (Tamang & Aravindan, 2017)	26 × 8 × 6mm (Tamang & Aravindan, 2017)	Epoxy resin and pure nickel slurry (Tamang & Aravindan, 2017)	Aluminum (Al) alloy Al6061-T6–Magnesium (Mg) alloy AZ31B (Tamang & Aravindan, 2017)	525 s (Tamang & Aravindan, 2017)
3.	900 W and 2.45 GHz (Satnam Singh et al., 2019)	20 × 10 × 10 mm (Satnam Singh et al., 2019)	Epoxy resin and EWAC powder slurry (Satnam Singh et al., 2019)	Cast iron-cast iron (Satnam Singh et al., 2019)	400 ± 20 s (Satnam Singh et al., 2019)
4.	900 W and 2.45 GHz (V. Kumar & Sehgal, 2020)	500 × 3.5 × 5mm (V. Kumar & Sehgal, 2020)	No interfacial material was used (V. Kumar & Sehgal, 2020)	Duplex stainless steel SS2205-duplex stainless steel SS2205 (V. Kumar & Sehgal, 2020)	300 s (V. Kumar & Sehgal, 2020)
5.	900 W and 2.45 GHz (S. Kumar et al., 2020)	18 × 3.5 × 5 mm (S. Kumar et al., 2020)	Stainless steel SS316 (S. Kumar et al., 2020)	Stainless steel SS304-stainless steel SS316 (S. Kumar et al., 2020)	360 s (S. Kumar et al., 2020)
6.	800 W and 2.45 GHz (Soni et al., 2018)	40 × 10 × 6 mm (Soni et al., 2018)	Nickel powder and Blumer 1400XX slurry (Soni et al., 2018)	Stainless steel SS316-stainless steel SS316 (Soni et al., 2018)	300 s (Soni et al., 2018)
7.	800 W (Bagha et al., 2019)	40 × 5 × 3 mm (Bagha et al., 2019)	No interfacial material was used (Bagha et al., 2019)	Stainless steel SS304-stainless steel SS304 (Bagha et al., 2019)	~300 s (Bagha et al., 2019)
8.	900 W and 2.45 GHz (Gamit et al., 2017)	Inner diameter: 1 mm, thickness: 2 mm, length: 20 mm (Gamit et al., 2017)	Nickel powder and epoxy resin slurry (Gamit et al., 2017)	Mild steel pipe-mild steel pipe (Gamit et al., 2017)	480 s (Gamit et al., 2017)
9.	900 W and 2.45 GHz (Satnam Singh et al., 2017)	25 × 10 × 6 mm (Satnam Singh et al., 2017)	EWAC powder and epoxy resin slurry (Satnam Singh et al., 2017)	Hastelloy-Hastelloy (Satnam Singh et al., 2017)	480 ± 10 s (Satnam Singh et al., 2017)
10.	900 W and 2.45 GHz (A. Kumar et al., 2019)	6 × 5 mm (A. Kumar et al., 2019)	No interfacial material was used (A. Kumar et al., 2019)	Stainless steel SS304-stainless steel SS316 (A. Kumar et al., 2019)	460 s (A. Kumar et al., 2019)

From different experimentation done by researchers, it was found that the exposure time required for joining different materials lies in the range of 5–15 minutes.

3.6 MECHANICAL AND METALLURGICAL CHARACTERISTICS OF JOINTS

After the formation of a joint, its mechanical and metallurgical properties are checked and compared with the base metal. These properties help to employ the various joining techniques for different materials in different applications based upon their characteristics. These properties further help to optimize the method and technique for further studies. Many researchers worked on joining materials and studied their various mechanical and metallurgical properties using standard techniques and methods. Metallurgical properties like microstructure, elemental composition, and chemical composition can be determined by scanning electron microscope, energy dispersive spectroscopy (EDS) and x-ray diffraction. Mechanical properties like microhardness and ultimate tensile strength can be determined by using Vicker's hardness test and a universal testing machine (UTM). Different joints formed by researchers are discussed in the following.

3.6.1 METALLURGICAL PROPERTIES

Srinath et al. (2011b) performed an experimental study on joining stainless steel (SS316) with mild steel with an exposure time of 450 s using a nickel-based interfacial powder. SEM analysis showed that the joint formed was homogeneous and the base metals were fused with each other, and there was complete melting of interface powder (i.e., Ni). EDS spectrum showed the presence of Ni and Fe in large concentrations; it also predicted the presence of chromium at the joint region. XRD analysis confirmed the presence of chromium carbide ($Cr_{23}C_6$) and cementite (Fe_3C) (Srinath et al., 2011b).

Srinath et al. (2011a) made a joint bulk copper using the MHH method using copper as interfacial material. The EDS spectra and XRD analysis are shown in Figure 3.8. Electron dispersive spectroscopy revealed that there was Cu in rich amounts, and O was present. XRD images showed the presence of Cu and CuO at the joint (Srinath et al., 2011a).

Bansal et al. (2014) did an experimental study on joining stainless steel (SS316) plates using stainless steel powder as the interfacial powder. The microstructure of the joined steel was studied using back scattered electron images. It showed the formation of carbides at grain boundaries. XRD images confirmed that there were various compounds at the joint: iron-nickel and chromium carbide (Bansal et al., 2014).

Singh et al. (2015) studied various properties and characteristics of aluminum joints formed through the MHH technique. SEM images showed the crack-free, homogeneous joining of metals. Also, x-ray diffraction spectra found the presence of Al and AlO at the joint region (Shivinder Singh et al., 2015).

Gamit et al. (2017) conducted experiments on joining mild steel pipes using MHH with nickel as interfacial material. After the joining of pipes, their metallurgical

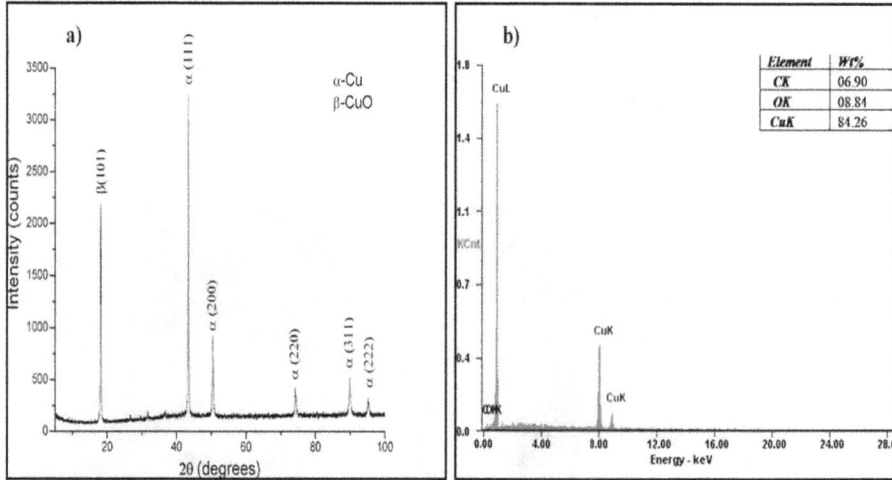

FIGURE 3.8 a) XRD analysis of copper joint; b) EDS spectra of joint formed (Srinath et al., 2011a).

Source: (Reproduced with the permission of rightsholder)

properties were studied. It was found through SEM (Figure 3.9) that the internal and external surfaces of the joint had a homogeneous and dense structure. It showed the micro-porosity of external part was greater compared to the internal joint part. XRD analysis showed nickel carbide intermetallic and iron (Fe) (Gamit et al., 2017).

Singh et al. (2018) (Satnam Singh et al., 2019) joined bulk cast iron using an MHH technique with EWAC powder as filler powder. Its various characteristics were studied. SEM showed smaller HAZ and cellular structure of grains were formed. It was found using EDS spectra that nickel was present at the joint surface and carbon, carbide, and chromium at grain boundaries (Satnam Singh et al., 2019).

Gupta, Kumar et al. (2021) formed a joint of mild steel (MS1020) using microwave technology and studied its various characteristics. Its metallurgical properties were investigated using XRD analysis, and it was found that Fe_3C, NiSi, and Fe_3Si were present (Gupta, Kumar, et al., 2021).

Gupta, Chaudhary et al. (2021) fabricated a brass joint using the MHH technique. SEM analysis (Figure 3.10) gave information on the microstructure of the joint. Cracks weren't present in the joint formed with uniform grain structure. XRD analysis confirmed the presence of Cu_2Zn, MnS, NiSi, and $NiSi_2$ (Gupta, Chaudhary, et al., 2021).

3.6.2 MECHANICAL PROPERTIES

Soni et al. (2018) fabricated stainless steel joint (SS316) using nickel nano-powder as a filler metal through MHH. Its mechanical properties were studied. Its tensile strength was investigated using UTM with maximum force of 100 kN. The ultimate

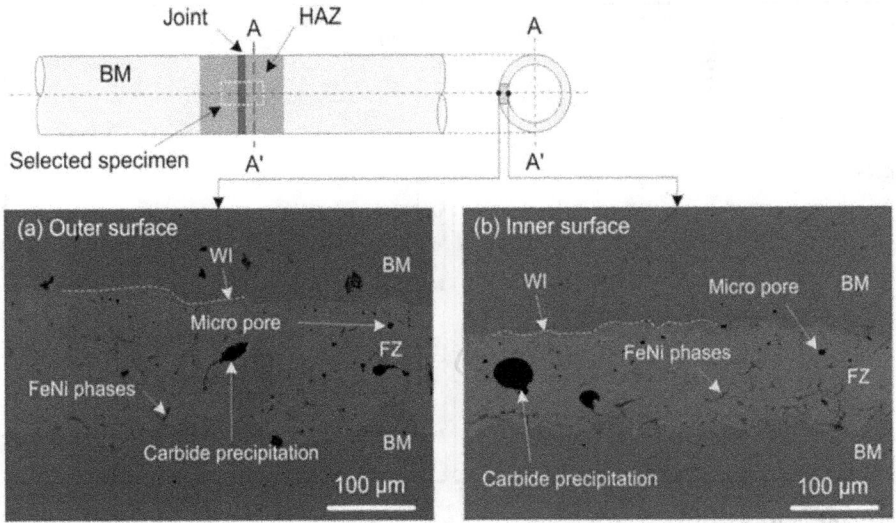

FIGURE 3.9 Typical SEM images of joint (a) outer surface; (b) inner surface (BM: base metal, WI: weld interface, FZ: fusion zone) (Gamit et al., 2017).

Source: (Reproduced with the permission of rightsholder)

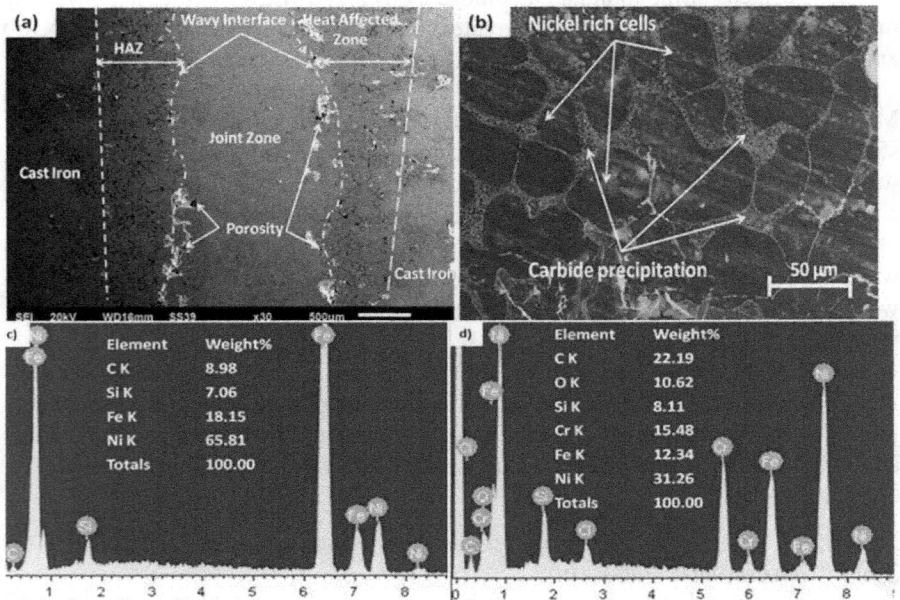

FIGURE 3.10 (a & b) Microstructure of brass joint (Gupta, Chaudhary, et al., 2021).

Source: (Reproduced with the permission of rightsholder)

tensile strength of the joined metal was 74% (385 MPa) of the bulk material (515 MPa. The hardness of the joint was investigated using Vicker's hardness testing machine and found to be 450 HV, which is more than the base metal owing to the formation of carbides from the graphite sheet used (Soni et al., 2018).

Bagha et al. (2019) joined stainless steel without any interfacial material in microwave radiation. It was found that the microhardness of the joint and the heat-affected zone were 260 and 200 HV, respectively, using Vicker's hardness checking apparatus. Also, its tensile strength was found using UTM at a load of 20 kN and was 271 MPa. Its elongation was found to be 2.34% (Bagha et al., 2019).

Kumar et al. (2020) performed experimentation on joining different grades of stainless steel using stainless steel powder as a filler material through microwave energy. Its microhardness was studied using Vicker's microhardness test. It showed the microhardness of the joint was 380 HV, which is comparatively higher than the base metal at ~200 HV (S. Kumar et al., 2020).

Kumar et al. (2020) (V. Kumar & Sehgal, 2020) joined duplex stainless steel without any filler metal in a domestic oven at frequency 2.45 Hz and 900 W power. The Vickers hardness of the joint was found under loading condition of 5 Kg. The microhardness of the joint formed was 62.7% (410 HV) more than that of the bulk material (252 HV) (V. Kumar & Sehgal, 2020).

Samyal et al. (2021) fabricated a lap joint of SS202 using microwave irradiation. Its characteristics were investigated at different exposure times. The microhardness of the joint region and HAZ were investigated with the application of 1 Kg load for 10s. It was found that the hardness of HAZ was unaffected by exposure time, which was nearly 250 HV. The hardness values for exposure times of 15, 17, 19, 21, and 25 min were 110, 125, 156, 232, and 479 HV, respectively. Increased microhardness of the joint with 25 min exposure time was a result of overheating, which leads to the formation of carbides (Samyal et al., 2021).

Singh et al. (2019) experimented with joining EN-08 using oxyacetylene gas welding, TIG welding, and MHH (EN-08 and epoxy resin slurry as interface material). It was found that the weld bead width of the microwave-processed joint was minimum, and there was minimum reduction in the hardness of the weld bead of the microwave-welded joint as compared to the other two welded joints. The MHH-processed joint showed the fewest defects owing to its uniform volumetric joining. Thus, the joint prepared through microwave heating is more efficient and defect free as compared to the oxyacetylene gas-welded joint and TIG-welded joint (Singh et al., 2019).

Handa et al. (2021) joined an iron-based alloy using graphite rods as susceptor material. The use of six graphite rods at the interface led to a concentration of heat at the joint, which helped in the formation of the joint. It was found that the Vickers hardness of the joint, 515 Hv, was more than that of bulk material (Handa et al., 2021).

3.7 LIMITATIONS AND RESEARCH GAPS

Most researchers have worked on joining using interface material with micro- or nano-sized powder, which is costly; less work is done without the use of any filler powder. In the sustainable development of joints, there is a need to optimize the technique, which is economical, too (Pal et al., 2021). The recent research has been carried out using domestic ovens, which can join only smaller pieces. Larger parts or components cannot be joined in domestic ovens. There is a need to work on using microwaves for joining larger parts and a wider range of parts for industrial applications (V. Kumar & Sehgal, 2021a).

3.8 CONCLUSIONS

This chapter discusses the joining of metals using sustainable energy resources. Most of the research has focused on different ways of using microwave energy (Salot et al., 2017). With time, methods are getting optimized. The MHH method has been worked out by most researchers to do experimentation. Though many metals have been joined using microwave energy, most of the work is done on joining stainless steel (Salot et al., 2017). Also, the interfacial material employed for joining the metals is nickel with a particle size of 40 μm. The time for which the joint is exposed to microwaves plays an important role in determining the characteristics of the joint. If the joint is exposed to microwaves for more than the optimal time, the metals start melting and fusing, which might result in blow holes and distortion of the joint (*Plasma Enhanced Microwave Joining*, n.d.). The conclusion can be made that joints formed by microwave irradiation are homogeneous and dense with a uniform grain size, and the hardness of the joined metals is greater than that of the bulk material due to the of the presence of carbon. The carbon present in the joint could be because of the susceptor powder and graphite sheet used to cover the joint, which gets mixed with the joint. The increased microhardness results in brittleness of the joined metals and reduced ductility.

REFERENCES

Agrawal, D. (n.d.). *Microwave Sintering, Brazing and Melting of Metallic Materials (Keynote)*.

Badiger, R. I., Narendranath, S., & Srinath, M. S. (2015). Joining of Inconel-625 alloy through microwave hybrid heating and its characterization. *Journal of Manufacturing Processes*, *18*, 117–123. https://doi.org/10.1016/j.jmapro.2015.02.002

Badiger, R. I., Narendranath, S., & Srinath, M. S. (2018a). Optimization of parameters influencing tensile strength of Inconel-625 welded joints developed through microwave hybrid heating. *Materials Today: Proceedings*, *5*(2), 7659–7667. https://doi.org/10.1016/j.matpr.2017.11.441

Badiger, R. I., Narendranath, S., & Srinath, M. S. (2018b). Optimization of process parameters by Taguchi grey relational analysis in joining Inconel-625 through microwave hybrid heating. *Metallography, Microstructure, and Analysis*, *8*(1), 92–108. https://doi.org/10.1007/S13632-018-0508-4

Bagha, L., Sehgal, S., Thakur, A., & Kumar, H. (2017). Effects of powder size of interface material on selective hybrid carbon microwave joining of SS304–SS304. *Journal of Manufacturing Processes*, *25*, 290–295. https://doi.org/10.1016/j.jmapro.2016.12.013

Bagha, L., Sehgal, S., Thakur, A., Kumar, H., & Goyal, D. (2019). Low cost joining of SS304-SS304 through microwave hybrid heating without filler-powder. *Engineering Research Express*, *1*(2). https://doi.org/10.1088/2631-8695/ab551d

Bansal, A., Sharma, A. K., Kumar, P., & Das, S. (2014). Characterization of bulk stainless steel joints developed through microwave hybrid heating. *Materials Characterization*, *91*, 34–41. https://doi.org/10.1016/j.matchar.2014.02.005

Bansal, A., Sharma, A. K., Kumar, P., & Das, S. (2015). Structure–property correlations in microwave joining of Inconel 718. *JOM*, *67*(9), 2087–2098. https://doi.org/10.1007/S11837-015-1523-4

Cai, J., Xie, X. M., Baoshun, L., Huang, X. X., Chen, T. G., & Guo, J. K. (1992). Microwave joining of Bi1.6Pb0.4Sr$_2$Ca $_2$Cu$_3$Ox superconductors. *Superconductor Science and Technology*, *5*(10), 599–601. https://doi.org/10.1088/0953-2048/5/10/007

Chandrasekaran, S., Basak, T., & Ramanathan, S. (2011). Experimental and theoretical investigation on microwave melting of metals. *Journal of Materials Processing Technology*, *211*(3), 482–487. https://doi.org/10.1016/J.JMATPROTEC.2010.11.001

Cozzi, A. D., Clark, D. E., & Ferber, M. K. (2008). *Microwave joining of high-purity alumina* (pp. 155–162). The American Ceramic Society. https://doi.org/10.1002/9780470314821.CH18

Gamit, D., Mishra, R. R., & Sharma, A. K. (2017). Joining of mild steel pipes using microwave hybrid heating at 2.45 GHz and joint characterization. *Journal of Manufacturing Processes*, *27*, 158–168. https://doi.org/10.1016/j.jmapro.2017.04.028

Gautam, U., & Vipin. (2021). Joining of metals using microwave energy. *Lecture Notes in Mechanical Engineering*, 1035–1039. https://doi.org/10.1007/978-981-15-5463-6_92

Gupta, P., Chaudhary, S., Kumar, V., Kumar, S., & Sahu, R. (2021). Investigation and characterization of brass joint fabricated by microwave hybrid heating process. *Materials Today: Proceedings*, *47*, 3973–3978. https://doi.org/10.1016/j.matpr.2021.04.054

Gupta, P., Kumar, V., Chaudhary, S., Kumar, S., Kumar Singh, J., & Kumar Chauhan, P. (2021). Investigation and characterization of microwave processed joint of MS1020. *Materials Today: Proceedings*, *47*, 3877–3883. https://doi.org/10.1016/j.matpr.2021.03.523

Handa, V., Goyal, P., & Sehgal, S. (2021). Development of novel experimental procedure for microwave hybrid heating based joining using graphite rods. *Materials Today: Proceedings*, *45*, 5769–5771. https://doi.org/10.1016/j.matpr.2021.02.601

Kamble, D. L., Kumar Sahu, R., Narendranath, S., & Badiger, R. I. (2021). Effect of input power and interfacial powder size on microwave joining of different materials: A review. *Materials Today: Proceedings*, *46*, 194–197. https://doi.org/10.1016/j.matpr.2020.07.351

Kumar, A., Sehgal, S., Singh, S., & Bagha, A. K. (2019). Joining of SS304-SS316 through novel microwave hybrid heating technique without filler material. *Materials Today: Proceedings*, *26*, 2502–2505. https://doi.org/10.1016/j.matpr.2020.02.532

Kumar, S., Sehgal, S., Singh, S., & Bagha, A. K. (2020). Investigations on material characterization of joints produced using microwave hybrid heating. *Materials Today: Proceedings*, *28*, 1319–1322. https://doi.org/10.1016/j.matpr.2020.04.588

Kumar, V., & Sehgal, S. (2020). Joining of duplex stainless steel through selective microwave hybrid heating technique without using filler material. *Materials Today: Proceedings*, *28*, 1314–1318. https://doi.org/10.1016/j.matpr.2020.04.509

Kumar, V., & Sehgal, S. (2021a). Microwave hybrid heating based optimized joining of SS304/SS316. *Materials and Manufacturing Processes*, *36*(13), 1554–1560. https://doi.org/10.1080/10426914.2020.1854469

Kumar, V., & Sehgal, S. (2021b). Use of metallic filler powders in microwave hybrid-heating based joining of metals and alloys. *Metal Powder Report*. https://doi.org/10.1016/J.MPRP.2021.02.044

Mishra, R. R., & Sharma, A. K. (2016). Microwave-material interaction phenomena: Heating mechanisms, challenges and opportunities in material processing. *Composites Part A: Applied Science and Manufacturing*, *81*, 78–97. https://doi.org/10.1016/J.COMPOSITESA.2015.10.035

Naik, S. R., Gadad, G. M., & Hebbale, A. M. (2021). Joining of dissimilar metals using microwave hybrid heating and tungsten inert gas welding—A review. *Materials Today: Proceedings, 46*, 2635–2640. https://doi.org/10.1016/j.matpr.2021.02.322

Nandwani, S., Vardhan, S., & Bagha, A. K. (2019). A literature review on the exposure time of microwave based welding of different materials. *Materials Today: Proceedings, 27*, 2526–2528. https://doi.org/10.1016/j.matpr.2019.10.056

Pal, M., Kumar, V., Sehgal, S., Kumar, H., Saxena, K. K., & Bagha, A. K. (2021). Microwave hybrid heating based optimized joining of SS304/SS316. *Materials and Manufacturing Processes, 36*(13), 1554–1560. https://doi.org/10.1080/10426914.2020.1854469

Pal, M., Sehgal, S., Kumar, H., & Goyal, D. (2020). Use of nickel filler powder in joining SS304–SS316 through microwave hybrid heating technique. *Metal Powder Report, 76*, 1–6. https://doi.org/10.1016/j.mprp.2020.10.001

Plasma Enhanced Microwave Joining. (n.d.). Retrieved January 30, 2022, from www.researchgate.net/publication/23813900_Plasma_Enhanced_Microwave_Joining

Sahota, D. S., Bansal, A., & Kumar, V. (2020). Application of microwave in welding of metallic materials—A review. *Materials Today: Proceedings, 43*, 466–470. https://doi.org/10.1016/j.matpr.2020.11.997

Salot, S., Sehgal, S., Pabla, B., & Kumar, H. (2017). Microwave joining of metals: A review. *Research Journal of Engineering and Technology, 8*(3), 282. https://doi.org/10.5958/2321-581X.2017.00048.4

Samyal, R., Bagha, A. K., & Bedi, R. (2019a). An experimental study to predict the exposure time for microwave based joining of different grades of stainless steel material. *Materials Today: Proceedings, 27*, 2449–2454. https://doi.org/10.1016/j.matpr.2019.09.217

Samyal, R., Bagha, A. K., & Bedi, R. (2019b). Microwave joining of similar/dissimilar metals and its characterizations: A review. *Materials Today: Proceedings, 26*, 423–433. https://doi.org/10.1016/j.matpr.2019.12.076

Samyal, R., Bagha, A. K., & Bedi, R. (2021). Evaluation of modal characteristics of SS202–SS202 lap joint produced using selective microwave hybrid heating. *Journal of Manufacturing Processes, 68*(PB), 1–13. https://doi.org/10.1016/j.jmapro.2021.07.018

Silberalitt, R., Ahmad, Iftikhar, Black, W. M., & Katz, J. D. (1993). Recent developments in microwave joining. *MRS Bulletin, 18*(11), 47–50. https://doi.org/10.1017/S0883769400038537

Silberglitt, R. (1995). *Microwave joining of SiC.* N.p. https://doi.org/10.2172/494132

Singh, G., Malhi, G. S., & Singh, T. (2019). Investigation of micro hardness and microstructure of microwave welded joint on EN-08. *A Journal of Composition Theory, 12*(7), 918–923. https://doi.org/19.18001.AJCT.2019.V12I7.19.10108

Singh, P., Prajapati, D. R., & Sehgal, S. (2021). A review on different types of material joints fabricated by microwave hybrid heating method. *Materials Today: Proceedings, 50*, 904–910. https://doi.org/10.1016/j.matpr.2021.06.253

Singh, S., Gupta, D., Jain, V., & Sharma, A. K. (2014). Microwave processing of materials and applications in manufacturing industries: A review. *Materials and Manufacturing Processes, 30*(1), 1–29. https://doi.org/10.1080/10426914.2014.952028

Singh, S., Singh, P., Gupta, D., Jain, V., Kumar, R., & Kaushal, S. (2019). Development and characterization of microwave processed cast iron joint. *Engineering Science and Technology, an International Journal, 22*(2), 569–577. https://doi.org/10.1016/j.jestch.2018.10.012

Singh, S., Singh, R., Gupta, D., & Jain, V. (2017). Preliminary metallurgical and mechanical investigations of microwave processed hastelloy joints. *Journal of Manufacturing Science and Engineering, Transactions of the ASME, 139*(6), 1–5. https://doi.org/10.1115/1.4035370

Singh, S., Suri, N. M., & Belokar, R. M. (2015). Characterization of joint developed by fusion of aluminum metal powder through microwave hybrid heating. *Materials Today: Proceedings*, *2*(4–5), 1340–1346. https://doi.org/10.1016/j.matpr.2015.07.052

Siores, E., & Do Rego, D. (1995). Microwave applications in materials joining. *Journal of Materials Processing Tech.*, *48*(1–4), 619–625. https://doi.org/10.1016/0924-0136(94)01701-2

Soni, P., Sehgal, S., Kumar, H., & Singh, A. P. (2018). Joining of SS316-SS316 through microwave hybrid heating by using nickel nano-powder. *International Journal of Applied Engineering Research*, *8*(8), 6446–6449.

Srinath, M. S., Murthy, P. S., Sharma, A. K., Kumar, P., & Kartikeyan, M. V. (2018). Simulation and analysis of microwave heating while joining bulk copper. *International Journal of Engineering, Science and Technology*, *4*(2), 152–158. https://doi.org/10.4314/ijest.v4i2.11

Srinath, M. S., Sharma, A. K., & Kumar, P. (2011a). A new approach to joining of bulk copper using microwave energy. *Materials and Design*, *32*(5), 2685–2694. https://doi.org/10.1016/j.matdes.2011.01.023

Srinath, M. S., Sharma, A. K., & Kumar, P. (2011b). Investigation on microstructural and mechanical properties of microwave processed dissimilar joints. *Journal of Manufacturing Processes*, *13*(2), 141–146. https://doi.org/10.1016/j.jmapro.2011.03.001

Tamang, S., & Aravindan, S. (2017). An investigation on joining of Al6061-T6 to AZ31B by microwave hybrid heating using active braze alloy as an interlayer. *Journal of Manufacturing Processes*, *28*, 94–100. https://doi.org/10.1016/j.jmapro.2017.05.027

Tamang, S., & Aravindan, S. (2019). Joining of Cu to SS304 by microwave hybrid heating with Ni as interlayer. *AMPERE 2019*. Polytechnic University of Valencia Congress. https://doi.org/10.4995/Ampere2019.2019.9813

4 Microwave Drilling
A Sustainable Drilling Method

Sarbjeet Kaushal

CONTENTS

4.1 INTRODUCTION

Drilling holes is the primary manufacturing operation in industries worldwide. Parts manufactured by forming, casting, and other manufacturing operations frequently require secondary machining operations such as milling, grinding, turning, and drilling. Drilling operation is most commonly used in the automotive, aerospace and aircraft industries. There are various kind of drilling methods used for creating holes in various kinds of materials. Mechanical drilling operation has been commonly used for this purpose; however, mechanical drilling is accompanied by noise, vibration, and emission of dust. Advanced drilling methods such as plasma arc and laser drilling use applications of thermal and ablation effects. Plasma arc drilling is extensively used to assemble different complex part profiles of different materials. On the other hand, laser drilling is widely used for drilling holes in hard and brittle materials. Both plasma and laser cutting are associated with higher installation costs. Further, water jet machining, abrasive jet machining, and ultrasonic machining are used for drilling holes for glass, ceramic, and so on. In recent decades, scientists and researchers have explored drilling methods that are environmentally friendly and energy efficient; in a nutshell, they are in search of sustainable methods of drilling. Microwave material processing has established itself as a environmentally friendly

DOI: 10.1201/9781003269298-4

TABLE 4.1

Comparison between Mechanical, Microwave, and Laser Drilling Operations

	Mechanical Drill	Microwave Drill	Laser-Based Drill
Principle	Grinding by mechanical friction	Melting (softening by microwave heating)	Evaporation by infrared radiation
Presence of mechanical contact	Yes	Yes	No
Presence of radiation	No	Localized (near-field)	Beam shape (far-field)
Pollution	Dust, noise	No	No
Cost	Low	Medium	High

Source: Eli Jerby and Dikhtyar (2003).

sustainable manufacturing technique that can process materials in less time without consuming too much power (B. Singh et al. 2018; Kaushal et al. 2017; Gupta et al. 2012; Mishra and Sharma 2016). Microwave material processing techniques have been used in sintering, joining, cladding, casting, and so on (Kaushal and Singh 2021; S. Singh et al. 2020; Agrawal 2006; Srinath et al. 2011). Owing to its unique advantage of volumetric heating, this process can provide lower thermal stresses in the material being processed as compared to conventional material processing techniques. Microwave drilling is a non-conventional machining process, which employs the microwave material interaction phenomenon for the drilling of materials. Material is drilled by converting microwave energy into thermal energy at the desired location, and the hole is drilled through the material ablation process. In 1999, Jerby and Dikhtyar (2000) filed a US patent on microwave drilling of non-conductive materials. Jerby et al. (2004) further reported a method of drilling nonconducting materials through the microwave drilling method. In this method, a local hot spot was generated through nearfield microwave radiation. A microwave source at 2.45 GHz and 1 KW was used with a directional coupler, a cascade with a microwave drill head. The authors successfully drilled holes in concrete, ceramics, basalt, silicon, and glass. A comparison between different drilling operations is illustrated in Table 4.1 (Eli Jerby and Dikhtyar 2003). In this chapter, the various findings reported in the domain of microwave drilling are presented.

4.2 MICROWAVE DRILLING OF CERAMIC-BASED MATERIALS

Jerby et al. (2003) reported a method of drilling hard non-conductive ceramic material using localized microwave radiation. The authors achieved drilling of ceramic materials using a portable and relatively simple drilling tool utilizing a conventional microwave source. Holes of 0.5 to 13 mm were produced using a drilling head with a coaxial feed with a near-field concentrator. This near-field concentrator focused microwave radiation on the selected region under the drilled material surface. This process is noise and dust free and does not require rotating parts.

Eli Jerby and Mark Thompson (2004) further successfully conducted drilling of a non-conductive ceramic-based thermal barrier coating (TBC) layer using the microwave drilling method. The experimental setup consisted of a 2.45-GHz magnetron coupled with an isolator, reflectometer, E-H tuner, and transition from rectangular guide wave to coaxial microwave drill (Figure 4.1). In this method, the authors utilized a coaxial open-ended microwave applicator, which induces localized thermal run-away in a dielectric material. In this technique, the microwave power density is concentrated on the contact point of the TBC layer, and a local hot-spot is produced. With the rise in temperature, the dielectric coupling of ceramic increases, and at the hot spot, the microwave absorption rate increases. This results in a rapid rise in the local temperature of ceramic to its melting point. A hole is created in this melting region by inserting the drill bit in it. It is shown in Figure 4.2(b) that a smooth circular exit hole is created using microwave drilling.

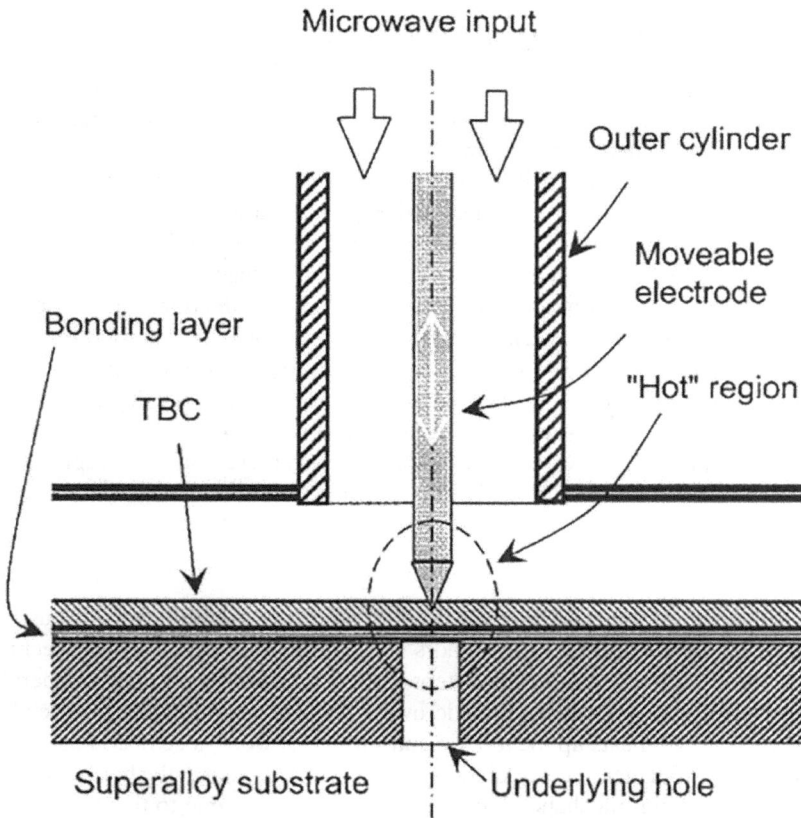

FIGURE 4.1 Schematic of the experimental setup for microwave drilling of thermal barrier coatings (Eli Jerby and Mark Thompson 2004).

Source: (Figure reused after permission of the publisher)

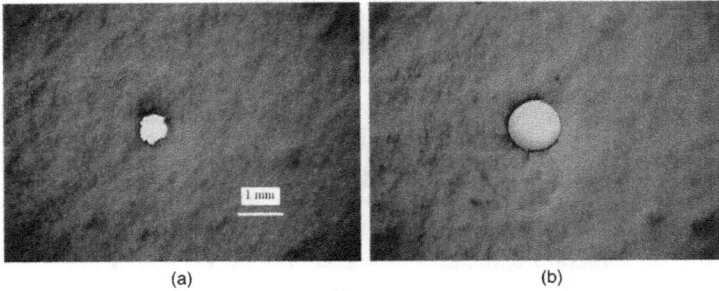

(a)　　　　　　　(b)

FIGURE 4.2　Optical macrograph: Top views before (a) and after (b) exposure of an underlying 1.27-mm-diameter hole through a dense TBC by the microwave drill (Eli Jerby and Mark Thompson 2004).

Source: (Figure reused after permission of the publisher)

4.3　MICROWAVE DRILLING OF BONES

The feasibility study of drilling in trabecular and cortical bone tissues was carried out through microwave radiation by (Eshet et al. 2006). The authors reported that the hot spots associated with microwave drilling have the potential to overcome the issues faced by mechanical drilling such as formation of debris resulting in foreign body reactions. When the microwave drill is passed through the bone's blood vessel, instead of rupturing the blood vessel, the hot spots contacting the blood vessel weld the vessel, which otherwise gets ruptured during mechanical drilling. Hence, microwave drilling of bones results in a lower risk of infection. Also, microwave drilling is quicker and geometrically precise. The authors also reported that Invitro microwave drilling in bones resulted in many encouraging results. Unlike mechanical drilling, the mechanical properties of the bones did not get degraded through microwave drilling. Microwave drilling produced smoother holes than mechanical drilling. Also, a minimal value of carbonization was present at the margin of holes in the case of microwave drilling.

4.4　MICROWAVE DRILLING OF GLASS

The mechanism of material removal through microwave drilling was discussed by Lautre et al. (2015a, 2015b). The authors utilized the microwave drilling technique for drilling soda lime glass. Various steps involved in microwave drilling operations are shown in Figure 4.3. During the microwave drilling operation, microwave plasma was formed, which heats up the tool tip and surface of workpiece in contact with the tool tip (Step I). Further, the workpiece material gets softened at the hotspot (Step II). The constrained plasma shape is formed in the workpiece due to the penetration of tool tip in the softened workpiece (Step III). This constrained plasma causes a high rate of melting and evaporation of workpiece materials in the vicinity of the tool tip (Step IV). Last, material ejection in the residue (molten work material) form at the entry and exit of hole is visible (Step V).

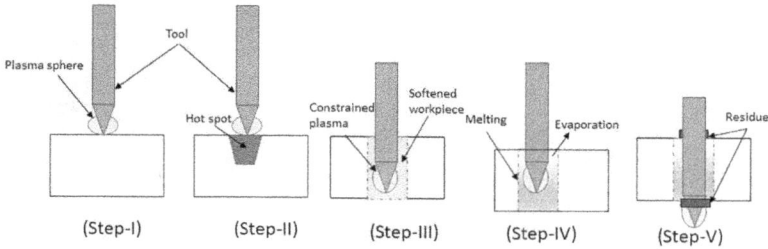

FIGURE 4.3 Various steps involved in microwave drilling of glass (Lautre et al. 2015a).

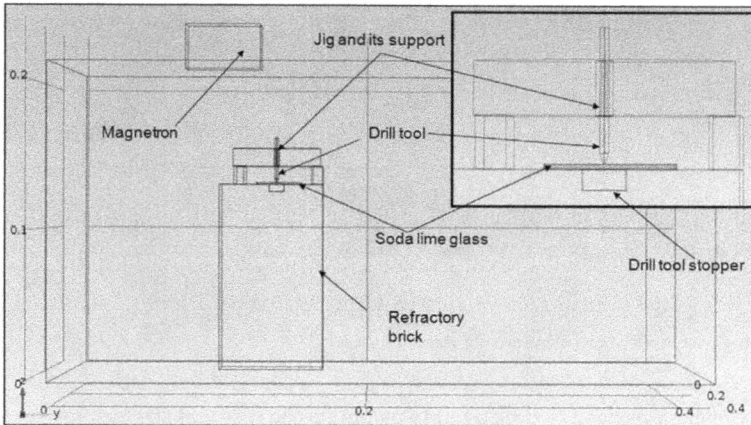

FIGURE 4.4 Schematic diagram showing setup for microwave drilling of glass (Lautre et al. 2015b).

Source: (Figure reused after permission of the publisher)

The microwave drilling of soda lime glass was carried out by Lautre et al. (2015b). Holes of approximately 900 μm diameter were drilled on soda lime glass using microwave energy at 2.45 GHz at a power range of 90–900 W in a multimode applicator. Material drilling was achieved through thermal ablation phenomenon with the creation of plasma by microwave energy through the tool. A nichrome wire drill bit was used as a tool and was held in a jig of 15.7 g dead weight. A refractory brick was used to hold the glass plate, and it also worked as a dummy load. The glass plate and drill bit were placed in the microwave applicator in such a way that the microwave energy concentration was maximum in a spatial position. During the microwave interaction, the microwave energy was concentrated at the tool tip due to its lower skin depth, the tip of the tool was heated up, and plasma was formed at the tool tip. In the vicinity of the tool tip, a localized microwave heating phenomenon was initiated, and the high heat of the plasma zone resulted in the melting of workpiece material underneath the tool tip. The tool tip is gravity fed, so the tool advances automatically in the material as the workpiece is thermally ablated. In this manner, the workpiece materials get drilled. The complete microwave drilling setup is shown in Figure 4.4.

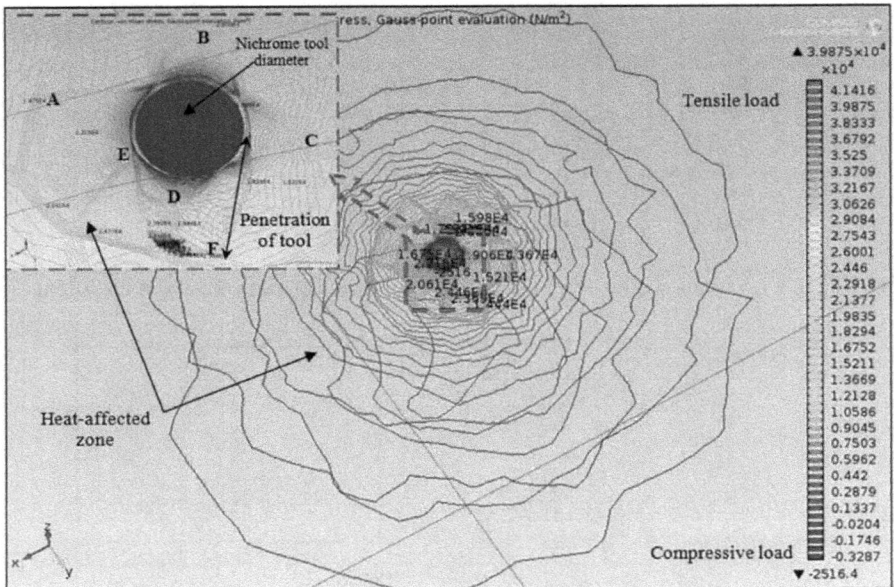

FIGURE 4.5 Stress variation on the surface during microwave drilling (Lautre et al. 2015b).

Source: (Figure reused after permission of the publisher)

A photoelastic approach was utilized to calculate the residual stresses induced in the drilled glass during microwave heating. The variation in the stresses on the surface of tool are shown in Figure 4.5, which reveals the presence of compressive stresses in the four quadrants of tools. The variation in the stresses near the hole boundary resulted in the generation of cracks in the heat-affected zone. It was concluded that the cracking of the glass can be controlled during microwave drilling using a suitable surface treatment.

4.5 ROLE OF VARIOUS PARAMETERS IN THE PERFORMANCE OF MICROWAVE DRILLING

4.5.1 ROLE OF DIELECTRIC MEDIA AND CONCENTRATOR MATERIALS ON MICROWAVE DRILLING

To reduce cracks and other thermal damage, in 2018, Kumar and Sharma reported the role of using different media in machining zones during microwave drilling of borosilicate glass. Different tool concentrator materials such as tungsten carbide, graphite, and stainless steel were used as tools/concentrators owing to their good thermal stability. Four liquids, palm oil, transformer oil, electric discharge machining (EDM) oil, and coconut oil were used as dielectric liquids in which workpieces were submerged. The machining results and hole qualities were measured in terms of heat-affected zone formed and overcut/taper, respectively. A microwave drilling setup diagram is illustrated in Figure 4.6. Various steps involved in microwave drilling are shown in Figure 4.7. In the first step, the tool concentrator is held on the surface of the workpiece [Figure 4.7(a)]. In the second step, the microwave field is allowed to enter in the applicator, and the electrons present on the surface of tool arrange themselves in such a way that the formation of electric field results and cancels out the applied electromagnetic field, which leads to an increase in the density of the surface charge [Figure 4.7(c)] at the tip of the concentrator, and formation of plasma occurs. This high-temperature plasma causes thermal ablation of material

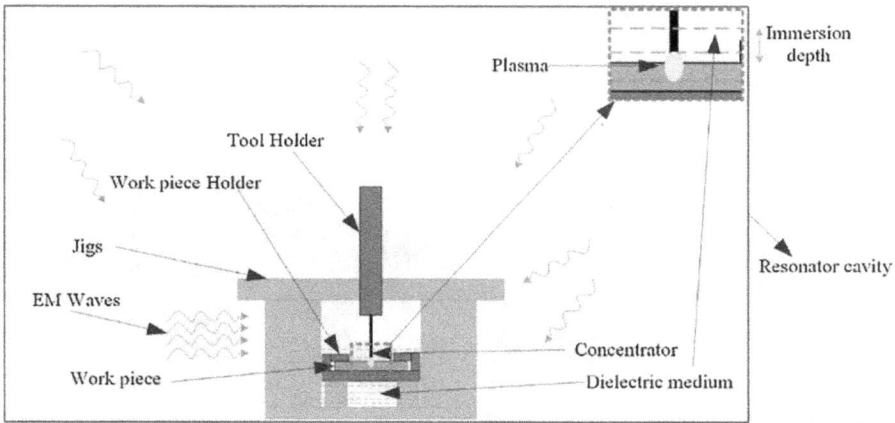

FIGURE 4.6 Setup of microwave drilling process (Kumar and Sharma 2018).

Source: (Figure reused after permission of the publisher)

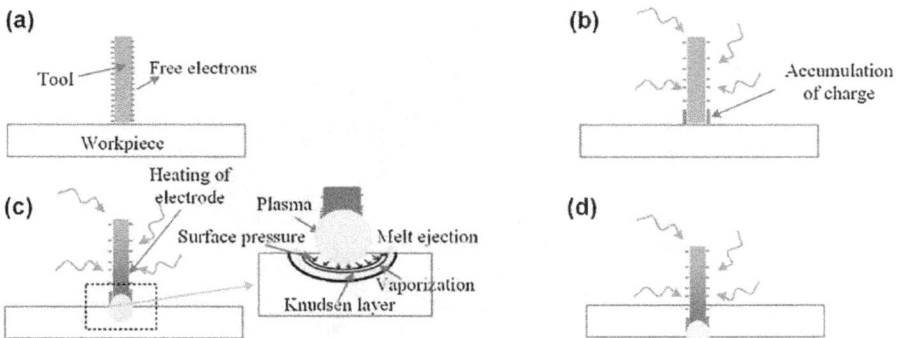

FIGURE 4.7 Various steps involved in microwave drilling: (a) free electrons present on toll surface; (b) accumulation of electrons around the circular edge of the concentrator after microwave setup was turned on; (c) formation of plasma and material removal mechanism; (d) Microwave drilling in progress (Kumar and Sharma 2018).

Source: (Figure reused after permission of the publisher)

from the workpiece surface. In this way, the material is removed beneath the concentrator tip, and due to gravity, the concentrator advances into the workpiece material; thus, microwave holes are drilled. It was observed that coconut oil and palm oil dielectric mediums resulted in fewer cracks around the microwave-drilled holes (Figure 4.8). It was reported by the authors that the high dielectric strength of both of these oils caused lower electric field strength around the concentrator tip than transformer oil and EDM oil.

4.5.2 Effect of Microwave Power on Drilling Time

Singh and Sharma (2019) reported the influence of microwave exposure power on drilling time. The authors used two microwave applicators, one with exposure power

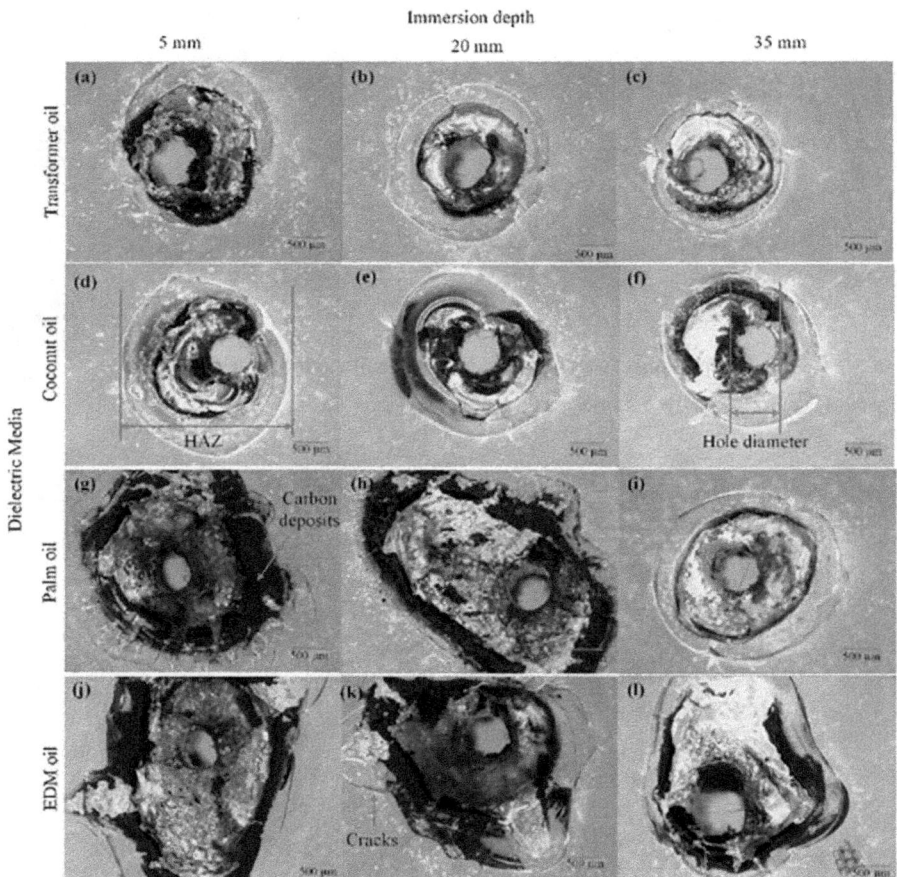

FIGURE 4.8 Typical images of the holes drilled using different dielectrics at different immersion depths with graphite concentrator (Kumar and Sharma 2018).

Source: (Figure reused after permission of the publisher)

0.7 kW and other with 3 kW, for experimental purposes to drill holes in a stainless steel sheet of 0.6 mm thickness. It was observed in the study that there was a reduction of 80% of the drilling time at 3 kW power in comparison to 0.7 kW power. The average hole diameter in the case of 0.7 kW microwave power was 1 mm, while in the case of 3 kW power, it was 0.6 mm. The smaller hole diameter in the case of higher power was attributed to the lower ablation time of the material in the case of 3 kW, due to which the plasma channel had less time to heat the surrounding material.

4.5.3 Role of Drill Bit Material in Microwave-Assisted Drilling

Lautre et al. (2014) reported the performance of microwave drilling process using various material bits in their research. A customized microwave oven working at 2.45 GHz was used as applicator in the present investigation. Perspex material was used as work material and was machined to dimensions 30 × 12 × 11 mm. Seven kind of drill bits, thin steel bit, thin copper bit, steel helical groove, mild steel flux coated, short tapered copper, long tapered copper drill, and litz copper wire, were used in the current investigation. The exposure power and exposure time in the customized microwave setup were maintained as 700 W and 300 s, respectively. It was reported in the investigation that the copper-based drill bit material was identified as the best combination material, then the steel-based drill bit material, as it exhibited less wear and better drilling capability than the steel drill bit material.

4.6   CONCLUSIONS

In this chapter, various findings related to microwave drilling of different materials were reported. The influence of various process parameters such as drill bit materials, microwave power, dielectric media, and tool concentrator materials were discussed. With an increase in microwave power, there was a significant reduction in drilling time. Copper-based drill bit materials performed best in microwave drilling. The use of dielectric media in microwave drilling methods resulted in the less thermal and residual stress in the microwave drilled material. In medical operations, microwave drilling is safer than mechanical drilling from an infection point of view. Microwave drilling can be further explored in the domain of drilling holes in composite materials.

REFERENCES

Agrawal, Dinesh. 2006. "Microwave Sintering of Ceramics, Composites and Metallic Materials, and Melting of Glasses." *Transactions of the Indian Ceramic Society* 65 (3): 129–144. https://doi.org/10.1080/0371750X.2006.11012292.

Eshet, Yael, Ronit Rachel Mann, Abby Anaton, Tomer Yacoby, Amit Gefen, and Eli Jerby. 2006. "Microwave Drilling of Bones." *IEEE Transactions on Biomedical Engineering* 53 (6): 1174–1182. https://doi.org/10.1109/TBME.2006.873562.

Gupta, Dheeraj, Prabhakar M. Bhovi, Apurbba Kumar Sharma, and Sushanta Dutta. 2012. "Development and Characterization of Microwave Composite Cladding." *Journal of Manufacturing Processes* 14 (3): 243–249. https://doi.org/10.1016/j.jmapro.2012.05.007.

Jerby, Eli, O. Aktushev, V. Dikhtyar, P. Livshits, A. Anaton, T. Yacoby, A. Flax, A. Inberg, and D. Armoni. 2004. "Microwave Drill Applications for Concrete, Glass and Silicon." *Proceedings of the 4th World Congress Microwave & Radio-Frequency Applications*, no. December: 156–165. https://www.semanticscholar.org/paper/MICROWAVE-DRILL-APPLICATIONS-FOR-CONCRETE-%2C-GLASS-Jerby-Aktushev/428c061ddd3639 5e6cd3e9da96ccb86ca4d425d2

Jerby, Eli, and Vladimir Dikhtyar. 2000. "Method and Device for Drilling, Cutting, Nailing, and Joining Solid Non-conductive Materials Using Microwave Radiation." United States Patent (19), no. 19. https://patents.google.com/patent/EP1147684B1/en

Jerby, Eli, and Vladimir Dikhtyar. 2003. "The Microwave Drill." *Water Well Journal* 57 (7): 32–34.

Jerby, Eli, and A. Mark Thompson. 2004. "Microwave Drilling of Ceramic Thermal-Barrier Coatings." *Journal of the American Ceramic Society* 87 (2): 308–310. https://doi.org/10.1111/j.1551-2916.2004.00308.x.

Jerby, Eli, V. Dikhtyar, and O. Aktushev. 2003. "Microwave Drill for Ceramics." *Ceramic Bulletin* 82: 1–23. https://www.eng.tau.ac.il/~jerby/microwave_drill/Microwave%20 drill%20-%20AcerS_02%20B.pdf

Kaushal, Sarbjeet, Dheeraj Gupta, and Hiralal Bhowmick. 2017. "Investigation of Dry Sliding Wear Behavior of Ni–SiC Microwave Cladding." *Journal of Tribology* 139 (4): 041603. https://doi.org/10.1115/1.4035147.

Kaushal, Sarbjeet, and Satnam Singh. 2021. "Processing Strategy for High Strength Ni-Based Hybrid Composite Clad on SS 316L Steel through Microwave Heating." *Proceedings of the Institution of Mechanical Engineers, Part B: Journal of Engineering Manufacture* 236 (3). https://doi.org/10.1177/09544054211021360.

Kumar, Gaurav, and Apurbba Kumar Sharma. 2018. "Role of Dielectric Fluid and Concentrator Material in Microwave Drilling of Borosilicate Glass." *Journal of Manufacturing Processes* 33 (May): 184–193. https://doi.org/10.1016/j.jmapro.2018.05.010.

Lautre, Nitin Kumar, Apurbba Kumar Sharma, Pradeep Kumar, and Shantanu Das. 2014. "Performance of Different Drill Bits in Microwave Assisted Drilling." *International Journal of Mechanical Engineering & Robotics Research* 1 (1): 23–29 http://www.ijmerr.com/show-134-660-1.html

Lautre, Nitin Kumar, Apurbba Kumar Sharma, Shantanu Das, and Pradeep Kumar. 2015a. "On Crack Control Strategy in Near-Field Microwave Drilling of Soda Lime Glass Using Precursors." *Journal of Thermal Science and Engineering Applications* 7 (4): 1–15. https://doi.org/10.1115/1.4030478.

Lautre, Nitin Kumar, Apurbba Kumar Sharma, Pradeep Kumar, and Shantanu Das. 2015b. "A Photoelasticity Approach for Characterization of Defects in Microwave Drilling of Soda Lime Glass." *Journal of Materials Processing Technology* 225: 151–161. https://doi.org/10.1016/j.jmatprotec.2015.05.026.

Mishra, Radha Raman, and Apurbba Kumar Sharma. 2016. "A Review of Research Trends in Microwave Processing of Metal-Based Materials and Opportunities in Microwave Metal Casting." *Critical Reviews in Solid State and Materials Sciences* 41 (3): 217–255. https://doi.org/10.1080/10408436.2016.1142421.

Singh, Anurag, and Apurbba Kumar Sharma. 2019. "Effect of Drilling Conditions on Microwave-Metal Discharge during Microwave Drilling of Stainless Steel." *Applied Mechanics and Materials* 895 (2012): 253–258. https://doi.org/10.4028/www.scientific.net/amm.895.253.

Singh, Bhupinder, Sarbjeet Kaushal, Dheeraj Gupta, and Hiralal Bhowmick. 2018. "On Development and Dry Sliding Wear Behavior of Microwave Processed Ni/Al$_2$O$_3$ Composite Clad." *Journal of Tribology* 140 (6): 061603. https://doi.org/10.1115/1.4039996.

Singh, Satnam, Dheeraj Gupta, and Vivek Jain. 2020. "Fabricating In Situ Powdered Nickel–Alumina Metal Matrix Composites Through Microwave Heating Process: A Sustainable Approach." *International Journal of Metalcasting* 15 (3). https://doi.org/10.1007/s40962-020-00536-w.

Srinath, M. S., Apurbba Kumar Sharma, and Pradeep Kumar. 2011. "Investigation on Microstructural and Mechanical Properties of Microwave Processed Dissimilar Joints." *Journal of Manufacturing Processes* 13 (2): 141–146. https://doi.org/10.1016/j.jmapro.2011.03.001.

5 Physical Vapor Deposition Coating Process in Biomedical Applications
An Overview

*Sivaprakasam Palani, Elias G. Michael,
Melaku Desta, Samson Mekbib Atnaw,
Ravi Banoth, Suresh Kolanji*

CONTENTS

DOI: 10.1201/9781003269298-5

5.1   INTRODUCTION

In recent years, surface engineering has rapidly grown as the market for improved materials has increased. It is an enabling technology that comprises surface treatment and deposition of thin films and coating. These applications can be used in sports technology, aeronautics, petroleum, food, mining, electronics, biomedical/orthopedics, dentistry, cancer therapy, and art industries. Thin films or coatings give the substrate material superior properties, such as resistance to high temperatures, anti-radiation, anti-corrosion, anti-wear, resistance to fatigue, antibacterial properties, bio-adhesion, and biocompatibility. The films or coatings have a much thinner thickness and smaller volume compared to the bulk substrate; however, they considered the key property of workpieces (Michelmore 2016). Surface engineering is the act of altering or coating the surface of a component in order to improve its qualities and critical role in tribology. Thin films and coatings' performance is critically dependent on their physical properties, which are determined by the film morphology and phase composition. The physical vapor deposition (PVD) coating process is mostly employed in biomedical applications because of its biocompatibility, high hardness, excellent corrosion resistance, and inertness without changing the biomechanical properties of the substrate, and the film layer is dense and uniform (Dong 2003; Nouri and Wen 2015). It is a coating process in which atoms or molecules are thermally evaporated from a source, proceed through the deposition chamber without colliding with residual gas particles, and condense on the substrate (Kopova et al. 2019; Choi et al. 2016). Stainless steels (316L), titanium and its alloys, NiTi alloys, Au-based materials, cobalt-based alloys, and Sn alloys are all used in biomedical applications (Nouri and Wen 2015). Titanium alloys are now the most desirable metallic biomaterials due to their increased resistance to corrosion under static conditions. When dynamic circumstances such as fretting, wear, fatigue, and chemically induced corrosion are present, these materials have inherent limitations (Antunes et al. 2009). These alloys have a poor reputation for tribological properties, and their continuing use is limited to their tribological applications (Ji et al. 1999).

Many kinds of research have been done to increase the biomechanical properties of biomaterials for a particular application; one way to change the surface of biomaterials is to put hard coatings on them. In recent years, diamond-like carbon (DLC) has emerged as an attractive material; the coating is made of only carbon and hydrogen. It is made up of sp2-bonded clusters of amorphous carbon that are linked together by a random network of sp3-bonded, atomic sites, and it is used to protect metallic components (Roy and Lee 2007). Because of its low friction and high hardness, it has good tribological properties in diverse settings (Ding et al. 2019). Because of its biocompatibility, resistance to corrosion and wear, low coefficient of friction (CoF), and high chemical inertness, it has been widely utilized in biomedical applications, and, depending on the deposition technique, it could be a robust insulator (Muthuraja et al. 2019; Shim et al. 2019; Ankha et al. 2019). A good biomaterial should have all of these qualities if it is going to be used in orthopedic, cardiovascular, contact lens, or dental applications. These features make it possible for the implant to last longer and not be rejected by the body's tissue next to it (Oliveira et al. 2014). Titanium and its alloys have a specific strength, excellent corrosion resistance, and 600 °C service temperature, giving them a wide range of uses in many applications such as engine turbines, ocean exploration, deepwater drilling, navigation, ocean exploration, aerospace, and the medical field. However, Ti-based biomaterials' weak tribological characteristics may cause greater wear and friction, limiting implant life (Dong et al. 2019; Khorasani et al. 2015). Premature failure owing to severe wear and corrosion is one of the major problems of titanium alloy implants. These often lead to a total revision arthroplasty and also expose the human body to toxic elements presented in the composition of implant alloy. A DLC film was applied to CP-Ti to provide a protective layer on the substrate surface, minimizing wear and erosion and increasing the implants' resistance to deformation (Khorasani et al. 2015). While oxygen has a significant influence in determining the strength range of many grades of titanium referred to as commercially pure titanium (CP-Ti), the phase stabilizes Ti alloys that are appropriate for biomedical applications (Khorasani et al. 2015).

Studies show the Ti-6Al-4V ELI is the most-used potential biomaterial, especially for biomedical implantation (Gepreel and Niinomi 2013; Abdelrhman et al. 2019). The Ti-6AL-4V ELI are used in dental implants, heart valves, joints, and bone implants as well as in the fabrication of several components. Due to its biocompatibility, wear resistance, good hardness and optimized CoF properties. Figure 5.1(a–d) shows metallic orthopedic devices for biomedical applications (Chen and Thouas 2015). This chapter looks at PVD processes currently used for wear protectors, including coating of ALTIN/DLC/TIALN, TIN, and so on. It also addresses the latest developments in surface-modification methods, various biomaterials, coating materials, and tribological properties.

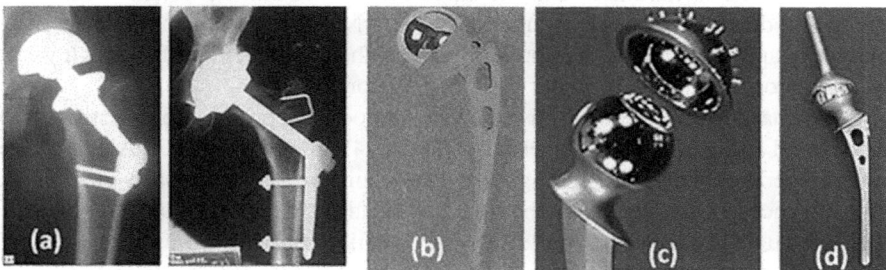

FIGURE 5.1 (a) Philip Wiles' metal-on-metal prosthesis made out of stainless steel; (b) Austin Moor's prosthesis made out of Co–Cr alloys;(c) McKee–Farrar prosthesis; (c) ring prosthesis; (d) Chanley's prosthesis.

5.2   METHODS OF DEPOSITION

A thin-film is a thin layer of material with a thickness varying between a few nanometers and a few micrometers. Characterizing the surface structure and physical properties of a new film is an important stage in developing commercial goods, and these attributes are closely tied to the deposition procedures. The deposition techniques are the most important and are usedby the surface coating technology to change the surface characteristics. The deposition process is regarded as a crucial component in the development of thin-film novel materials to address the growing demand from industries for multifunctional and multi-dynamic materials. Not all deposition techniques give similar properties like surface morphology, biocompatibility, tribological, hardness, optical, and corrosion (Ma et al. 2009). Generally, there are two commonly used methods of deposition techniques, physical and chemical depositions. Table 5.1 shows a comparison of various coating processes and their applications.

5.3   PHYSICAL VAPOR DEPOSITION

Physical vapor deposition is the least harmful to the environment. The PVD technique generates no hazardous waste in comparison to other coating processes like as electroplating or painting. PVD has a relatively low risk of releasing harmful materials into the environment, especially when compared to electroplating. PVD coatings, by themselves, contribute to environmental conservation. PVD coatings also contribute to product longevity, which benefits the environment in a variety of ways, including reduced landfill volume. PVD-coated products have a longer lifespan and are less likely to corrode and leach pollutants into the environment.

Ming'e et al. (2016) investigated TiN coatings deposited to a GCr_5 substrate by using different methods of PVD coating with a medium-frequency magnetron, cathode arc deposition, sputtering, and their composite methods by varying temperature, pulse bias, and deposition time to compare the mechanical and morphological properties of TiN coatings for cutting tool applications. X-ray diffraction, scanning electron microscope (SEM), scratch tests, friction, and wear tests characterize the coating to get the best coating methods. The results show that better extensive property like cohesion, hardness, and the smoother surface can be obtained the wear rate also reduced. Finally, they suggested that TiN thick coatings made by a PVD composite approach combining sputtering and cathode arc methods have high wear-resistant properties working under high loading conditions.

Gangatharan et al. (2016) used the PVD method of deposition to investigate the tribological and mechanical properties of Ti-6Al-4V coated with DLC/AlCrN at the high-temperature condition for lightweight automotive and aerospace vital parts. The influence of DLC/AlCrN on the wear and surface characterization of Ti-6Al-4V was also studied under a pin on disc wear machine, SEM, mini-universal testing machine, and Vickers nanoindentor deposited by the arc evaporation method with the consideration of operating temperature and time of deposition as parameters. The results reveal that there is no particle aggregation of particles detected using SEM during coating. DLC/AlCrN-coated titanium Ti-6Al-4V deposited by PVD has high

TABLE 5.1

Summary of Coating Methods and Their Applications

Method	Coating Quality	Coating Process rate	Bond Strength	Toxicology Analysis and Biological Properties	Application	Reference
Plasma immersion ion implantation	Thin injection layer inert to the surface, high biocompatibility	Not available	Moderate	Good biocompatible properties	Used for large, heavy, and complex-shaped implant surface	Yu et al. 2016 Shanaghi and Chu 2019
Plasma immersion ion implantation and deposition	Composition control Improved density and adhesion Suitable for 3D and complex surfaces	30–40 nm/min	Very high	Rapid osseointegration and biomechanical stability	Used for precision parts with wear resistance	Yang et al. 2007 Yu et al. 2016
Physical vapor deposition	Highest bonding force, uniform and dense layer formation	≈25–1000 nm min-1 in thickness	Very good	Excellent biocompatibility and osteointegration	Utilized for application of nitride coating on metallic elements	Behera et al. 2020 Liu et al. 2020
Chemical vapor deposition	Good control over the film's content and attributes, as well as flexibility	At high temperatures, a chemical process is used to deposit the material	-	It has a beneficial impact on osteoblast-like cell proliferation and activity, as well as sterilizing efficiency	Used for covering interior holes and intricate workpieces	Du et al. 2016 Youn et al. 2019
Sol gel	Easy to make a homogeneous multi-component oxide film, has excellent composition and microstructure control	Preparation of the sol-gel, transfer of the sol-gel to the substrate, aging, and drying are all steps in the process	3–55 MPa	Simulated Body Fluid (SBF) produces bone-like apatite and has high biological activity and antifungal properties	Coated implants are used in animals bones	El hadad et al. 2020 Ziabka et al. 2020

hardness, a high CoF, and better wear resistance than uncoated Ti-6Al-4V. There are different methods of PVD, and the most common are the following.

5.3.1 THERMAL EVAPORATION

Thermal evaporation is a well-known technique for covering a thin layer in which, due to high-temperature heating, the source material evaporates in a vacuum, allowing the vapor particles to pass and directly obtain a substrate where these vapors transform to a solid state again. In a vacuum setting, solid materials could evaporate by resistive heat to form a thin layer. In the form of a charge-holding boat or resistive coil, the mechanism requires the use of a powder or solid bar. The resistive boat/coil is subjected to a broad direct current (DC) to achieve the high melting points required for metals, where the high vacuum (below 1024 Pa) facilitates the evaporation of the metal and further carries it to the substrate. This method is specifically applicable to materials with low melting points. Figure 5.2 presents a schematic representation of an evaporation system.

5.3.2 ELECTRON BEAM EVAPORATION

It is one type of PVD in which an electron beam emitted by a charged tungsten filament under a high vacuum is bombarded with a target anode. The electron beam induces atoms to enable phase transition from the target to the vapor form. Due

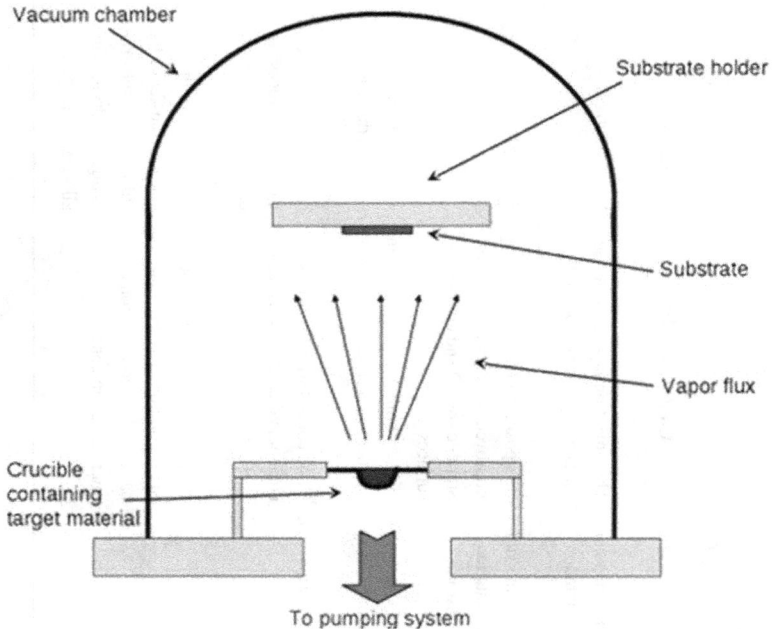

FIGURE 5.2 Schematic representation of evaporation (Martín-Palma and Lakhtakia 2013).

to their biological and mechanical properties, biomedical implants like orthopedics and dental implants can fail. Electron beam (E-beam) evaporation is an appropriate deposition technique to solve such problems. Gnanvel et al. (2018) used E-beam evaporation to deposit hydroxyapatite (HA) on Ti-13Nb-13Zr by aiming to improve mechanical and biological properties. The experimental finding shows that the resistance to oxidation of the HA-coated alloy was considerably greater than that of uncoated alloys.

5.3.3 ARC EVAPORATION

Arc evaporation/melting is a coating method that uses an arc (a discharge of electricity between two electrodes), shown in Figure 5.3. There is no evaporation in this type of deposition mechanism, but melting of the metals forms the necessary alloy due to the heat. The substrate is subjected to intense metallic ion bombardment, resulting in a clean and heated surface without any modifications in the properties of the substance of the coating.

The combination of a high substrate temperature and high-energy ions results in coats with outstanding adherence and density; however, drops flung from the target result in a high coated surface roughness when compared to the other active techniques.

5.3.4 SPUTTERING TECHNIQUES

The sputtering technique is often used by manipulating metal and oxide films for depositing crystalline structure and roughness of surfaces. The sputtering mechanism is based on the collision of the ions emitted from the discharge into the cathode

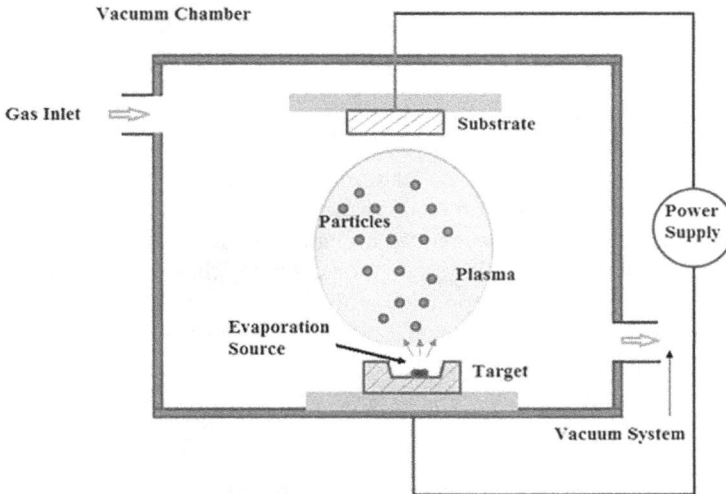

FIGURE 5.3 Schematic representation of evaporation.

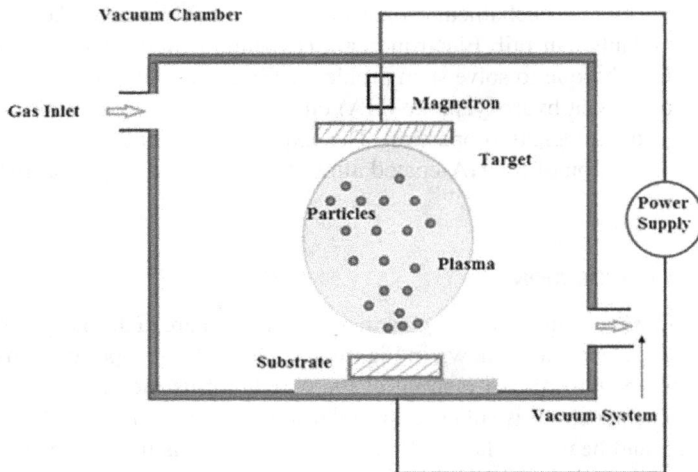

FIGURE 5.4   Schematic representation of sputtering.

molecules, leading to a higher kinetic energy release of the molecules from the cathode. Direct current and radio frequency (RF) sputtering are the two most common types of sputtering procedures. The first depends on DC power, which is usually used for target materials that are electrically conductive as a low-cost alternative that is easy to manage. RF sputtering uses RF power for most dielectric materials. A schematic representation of the PVD sputtering process is shown in Figure 5.4.

5.4 BIOMATERIALS

A biomaterial is an unviable product (capable of working effectively after implantation) designed to communicate with biological systems. With the effective and durable characteristics of biomaterials, their use is possible within a physiological medium. A suitable combination of biological, chemical, mechanical, and physical properties for the design of these characteristic features is given. Specifically, biomaterials are designed using material classes: metals, ceramics, composite, and polymer. Most of the biomaterials that are available today are either manufactured individually or in conjunction with the materials in these groups. Among these, metallic biomaterials are mostly used for biomedical applications due to their superior properties.

Biomaterials, often known as implantable biomedical devices, are natural or synthetically made artificial materials that are utilized as implant surfaces to substitute for the function of a missing or sick biological component. Artificial stents in blood veins, artificial heart valves, shoulder replacement, knee joints, elbows, ears, orodental structures, and protheses are only a few examples of the application area for biomaterials (Jammalamadaka and Tappa 2018). Metals, ceramics, polymers, composites, and their mixtures are all types of material utilized during manufacturing of biomaterials (Jayaraj and Pius 2018). Metal biomaterials are the oldest and most commonly utilized biomaterials in clinical practice, with good results. Stainless steel, cobalt, and titanium, as well as their alloys, are well-known metal biomaterials (Wilson 2018). Titanium and its alloys have exceptional mechanical qualities, such as lower elasticity modulus, better tensile strength, maximum toughness, and stronger fatigue resistance, when compared to other metals (Nasker and Sinha 2018). Moreover, due to its oxidized film passivation stability, titanium alloy has excellent corrosion resistance to physiological fluids with much better biocompatibility than other metallic biomaterials. Furthermore, titanium alloy's biological surface responses, particularly bioactivity and osseointegration, are favorable for clinical use (Liu et al. 2004).

Pure titanium and extra-low interstitial (ELI) Ti-6Al-4V alloy are two of the most-used biological alloys for orthopedic implants (Nasker and Sinha 2018; Im 2020; Kaur and Singh 2019). In particular, the Ti-6Al-4V alloy possesses a variety of properties that make it appropriate for surgical implants, including a low elastic modulus, sufficient tensile and fatigue strengths, and biocompatibility. Despite the high mechanical capabilities of titanium-based alloys, the exposed surface of titanium-based implants is recognized for its poor tribological qualities (Raval et al. 2019). Various problems have been observed that have limited the long-term usage of Ti-based implants, all of which have a significant effect on patient health and healthcare expenditures. Due to wear and corrosion, Ti-based implants, particularly Ti-6Al-4V alloys, leak metallic particles and ions into surrounding tissues. Metallic debris that is not excreted from the tissue causes health issues such as bone loss as a result of inflammatory reaction. Another major worry is the toxicity of liberated aluminum and vanadium ions, which are the primary cause of peripheral neuropathy, osteocalcin, and Alzheimer's disease in the long run (Bernhardt et al. 2021). While titanium has good biomechanical qualities, most titanium implants fail after 10–15 years, necessitating re-surgery. To minimize certain biomechanical and biological function failure, it is vital to increase the quality and dependability of Ti-based implants. The titanium alloy's surface must be changed to meet standard bio-medical application criteria in order to minimize harm to the human body.

Over the last 10 decades, many approaches to Ti-based implant surface alteration techniques have been used to improve wear characteristics, raise bone-to-implant contact ratio, and extend the life lifetime of Ti implants (Jin 2018). During the manufacturing process, the top surface of titanium implants will become polluted and oxidized. Such a surface layer has properties like being unevenly plastically deformed, strained unevenly, and a microstructural nature that is relatively inefficient (Liu et al. 2019). Such surfaces, it has been noticed, clearly limit the use of biomedical devices, and some surface changes are required to fulfill the lower standards. Another significant principle behind surface treatments is that they are used to achieve specific surface properties that differ from bulk materials. As a result, surface adjustments are required in order to address these issues. The best surface modification options improve explicit surface qualities while also preserving critical bulk material features (Antunes and de Oliveira 2015). Currently, a variety of surface modification approaches have been used to address various clinical issues and demands. Great mechanical alignment between the implant and the adjoining tissue, higher conductivity and inductivity (especially for bone implants), excellent tribological features, enhanced biocompatibility, and faster healing capability are all advantages of these surface-altering processes (Dobrzański 2015).

5.4.1 METALLIC BIOMATERIALS

For a long time, metallic implant materials have had enormous therapeutic value in the medical profession. SS 316L, CoCr-alloys, titanium and alloys (Cp-Ti, Ti-6Al-4V), zirconium niobium, aluminum alloys, and tungsten alloys are among the metals and metal alloys utilized for medical applications (Nouri and Wen 2015). Biomaterials, however, should also demonstrate high biocompatibility to be safely incorporated into the living organism, which ensures that they do not cause any adverse tissue reactions such as acute and chronic inflammation or irritation of the underlying tissues. Even metals with good corrosion resistance do not demonstrate optimum chemical stability in the very challenging environment of the human body, despite good mechanical qualities. As a result, metallic biomaterials emit toxic byproducts such as metal ions, which might result in a negative biological degradation response. The degradation products released not only accumulate in the tissues and organs nearby it, they also induce inflammation and allergic reactions. A post-implantation infection can also occur, requiring the implant to be revised or removed.

5.4.2 STAINLESS STEEL

Stainless steel is commonly found in orthopedics, joint prostheses, surgical and dental devices, plates and screws (bone), pins, rods, and coronary stents for fracture repair. It is appropriate for various applications, particularly for fracture fixation, due to its biocompatibility, strength, ductility, and working hardness.

A recent popular application for joint replacements such as hips and knees or spinal devices is not seen due to it slow strength and wear characteristics as compared to titanium and Co-Cr alloys, such as osseointegration and reduced MRI artifacts. Titanium and its alloys could be used to repair fractures in place of austenitic stainless steel alloys. However, the numerous possible attributes, the fact that fixation devices are not meant to be bone replacement devices, and lower cost to the patient have resulted in the widespread use of these alloys for previously delineated applications. There are several different stainless steel styles. Form 316 is an austenitic stainless steel-containing Cr-Ni-Mo. SS 316L is produced with various quantities of Fe, Cr, and Ni, with lower carbon quantities added. Figure 5.5 shows 316L stainless steel bone plates used for (a) humeral fracture and (b) tibial fracture (Li et al. 2020)

5.4.3 COBALT-CHROMIUM ALLOYS

Alloys of cobalt-chromium have high specific strength and are strong, durable, biocompatible metals that are resistant to corrosion. Co-Cr is a metal that is frequently used in knee implants, along with titanium. Similar to SS, the strong mechanical properties of Co-Cr alloys are due to the multiphase structure and carbide precipitation, which increases the hardness of Co-Cr alloys. In knee implants, these alloys are among the most commonly used metals. When high stiffness or a highly polished

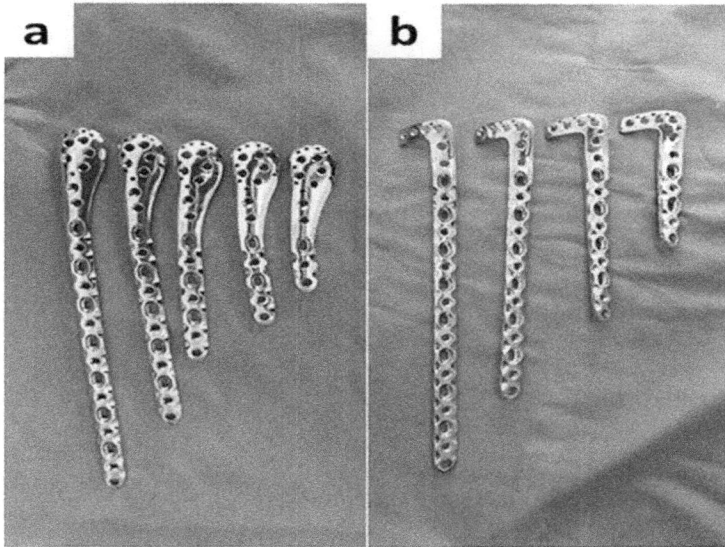

FIGURE 5.5   Stainless steel bone plates (a) humeral fracture and (b) tibial fracture (Li et al. 2020).

and extremely wear-resistant material is needed, Co-Cr alloys are particularly useful. They are usually made of Co with Cr, Mo, and traces of other elements in orthopedic implants. Although the incidence of allergic reactions related to the use of Co-Cr alloy is extremely rare, one area of worry is the possibility of metal ions being released into the body as a result of joint mobility. The biomedical applications of Co-Cr alloys in the hip stem, hip joint head, metal/metal articular joint, mechanical house valve housing, and vascular stent are shown in Figure 5.6.

5.4.4 TITANIUM AND ITS ALLOYS

Titanium (Ti) alloys exhibit extraordinary mechanical properties, including a high strength-to-weight ratio, outstanding corrosion resistance, and superior biocompatibility. Microstructures of Ti-alloy depend on the type of stabilizers present. α-stabilizers contain Al, O, N, and C; β-stabilizers contain Mo, V, Nb, Fe, and Cr; and neutrals contain Zr. Ti-6Al-4V is the most widely used Ti-alloy. Apart from their Young's modulus, all mechanical characteristics of titanium alloys are directly influenced by their alloy composition (Chen and Thouas 2015).

Commercial pure (CP) titanium (98.9–99.6% of Ti) is mainly α-titanium with a range of mechanical properties due to the difference in interstitial and impurity levels (Chen and Thouas 2015). When compared to α-β and β alloys of titanium, pure α-titanium exhibits better formability and excellent corrosion resistance characteristics. Hence, they are used in the food and petrochemical industries (Sandlöbes et al. 2019). CP titaniums are known for their excellent corrosion resistance and

(a) Hip stem

(b) Hip joint head

(c) Metal/metal articular joint

(d) Housing of a mechanical heart valve

(e) Vascular stent

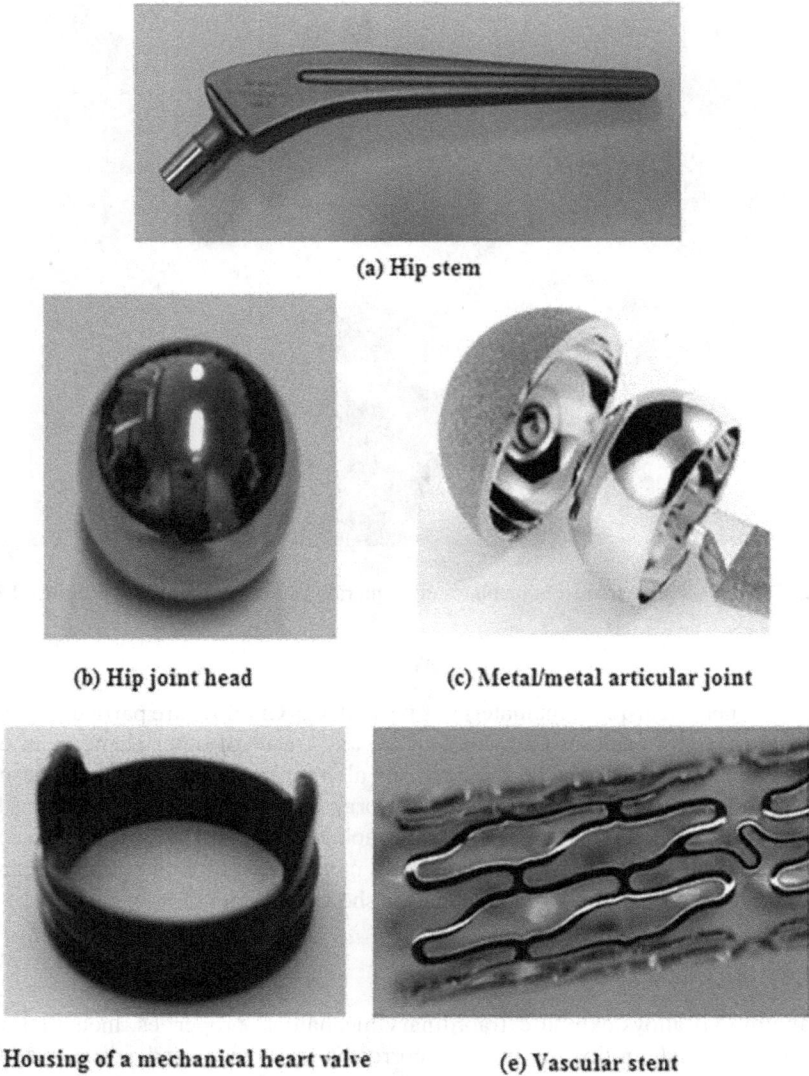

FIGURE 5.6 Biomedical applications of Co-Cr alloys (Tanzi 2019).

biocompatibility. Some disadvantages of these alloys include poor strength at ambient environment, and heat treatment has no effect on them. Medical applications of the alloy include dental implants, pacemaker cases, screws, and staples for spinal surgery. Some of the examples of α-titanium alloy include all ASTM Grade 1, 2, 3, and 4 alloys (Chen and Thouas 2015).

The other types of Ti-alloys are near-α or super-α alloys. They contain a slight amount of β stabilizers, not more than 5% (Polyakova et al. 2017). The characteristics of low strength at ambient temperature have limited the application of α or near-α Ti-alloys (Chen and Thouas 2015). The α-β microstructure can be strengthened by solution treating and aging (heat treatment). It will enhance the strength of the alloy up to 50%.α-β-Ti-alloys that can be used for orthopedic implants include Ti-6Al-4V, Ti-6Al-7Nb, Ti-5Al-2.5Fe, and Ti-3Al-2.5V. Figure 5.7 depicts titanium and its alloys in medical devices.

Application areas of Ti-6Al-4V and Ti-6Al-4V ELI (interstitial elements such as oxygen and iron are kept controlled to improve fracture toughness and ductility) include total joint replacement arthroplasty for hips and joints, whereas Ti-6Al-7Nb

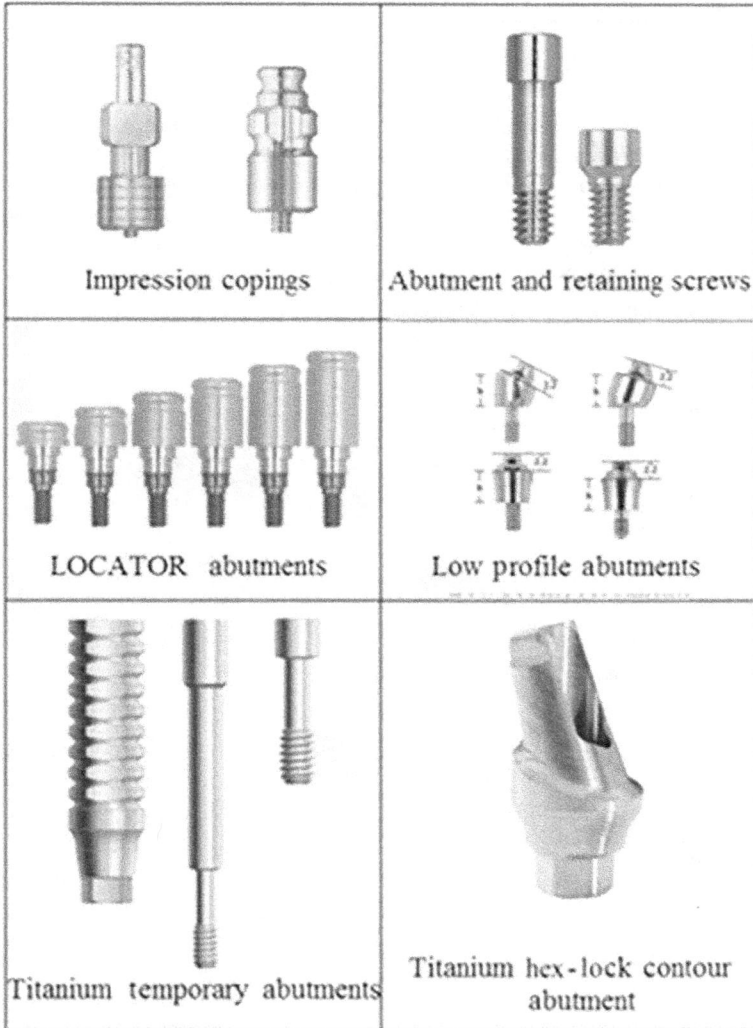

FIGURE 5.7 Titanium and its alloys in application (Gubbi and Wojtisek 2018).

can be used for femoral hip stems and fracture fixation plates (Chen and Thouas 2015); Yang (2017). β microstructure Ti-alloys have better strength compared to α-β-Ti-alloys [20]. They are commonly used as biomedical implant materials (Sandlöbes et al. 2019). Urena et al. (2019) highlighted the increasing studies of these alloys for knee and hip implants despite their poor tribological behavior. These alloys contain alloying elements such as zirconium, tantalum, and iron. Ti-13Nb-13Zr, Ti-Zr-Nb-Ta, and Ti-12Mo-6Zr-2Fe (TMZF) are some examples of β-microstructure Ti-alloys that can be used for orthopedic implants. High density, poor wear resistance, and low creep strength are disadvantages associated with these types of Ti-alloys.

5.4.5 NICKEL-TITANIUM ALLOY (NITINOL)

Because of their exceptional mechanical properties, nickel-titanium (NiTi) alloys have been a groundbreaking breakthrough and have found wide application in orthodontics, often as arch wires in the straight-wire technique. Nitinol, a group of Ti-Ni alloys that are nearly equiatomic, is commonly known and approved for medical use. The shape memory, super elasticity, and high wear resistance of Nitinol have made it possible to design novel instrumentation and implants in surgical fields ranging from orthopedics to vascular interventions. Nitinol's remarkable malleability and ductility enable it to be manufactured in a number of medical applications, such as wires, plates, bars, and tapes, shown in Figure 5.8.

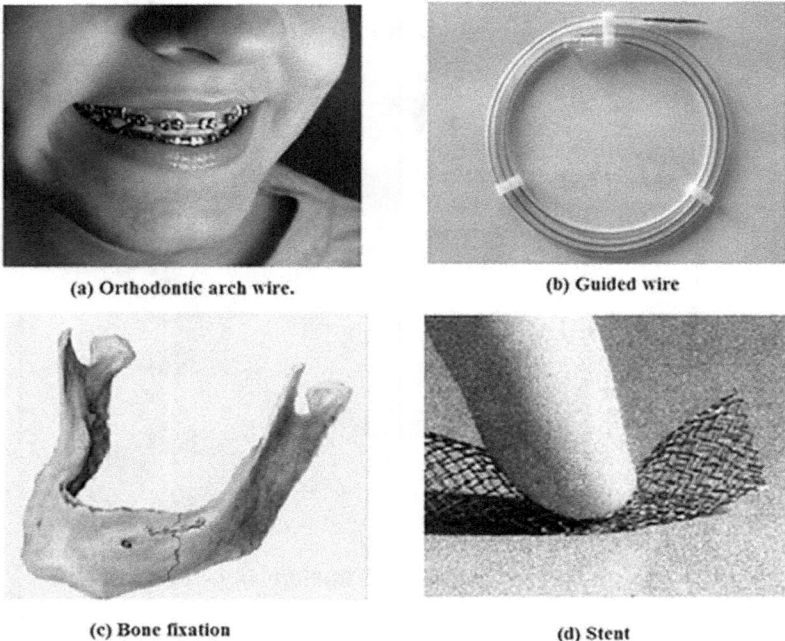

(a) Orthodontic arch wire.   (b) Guided wire

(c) Bone fixation   (d) Stent

FIGURE 5.8 Biomedical applications of Nitinol (Wadood 2016).

5.5 IMPLANT SURFACE AND MATERIAL PROPERTIES

The relationship between surface topography and material adhesion properties is critical in determining implantation success (Kheradmandfard 2018). As a result, there are attributes that define the surface properties of materials to determine if they are suitable for biomedical application, such as biocompatibility, mechanical properties, tribological features, and osseointegration, to name a few (Liang et al. 2018). The bulk properties of biomaterials, such as non-toxicity, fatigue strength, modulus of elasticity, and corrosion resistance, have long been recognized as critical in selecting the optimal biomaterials for a given biomedical application (Tüten et al. 2019). Interactions between the physiological environment and artificial material surfaces, the commencement of biological reactions, and the body's specific response courses are all events that occur after implantation. The biological environment's response to artificial medical devices is strongly influenced by the material surface. The main properties considered during selection of biomaterial implant are discussed in the following.

5.5.1 BIOCOMPATIBILITY

The word "biocompatibility" refers to the biological requirements for successful biomaterial application in the medical field, as well as the downside of excessive tissue responses. In fact, according to Ratner (2019), a material's potential to generate consecutive reactions from tissues following the implant procedure is dependent on both the material and the material degradation that occurs in the host's body. The host's reaction reflects the responses elicited by a system as a result of the presence of new data. Biodegradable materials, on the other hand, degrade quickly in the host body, which is an important feature to look for in biomaterials (Sommer et al. 2020).

5.5.2 MECHANICAL PROPERTIES

Prior to implementation, mechanical qualities are incorporated into biomaterial design. In the case of orthopedics, a lower Young's modulus is preferred to reduce the negative consequences of the stress shielding effect. As a result, fatigue resistance and strength are critical in the implementation of a cap and a stent (Veerachamy et al. 2018). For implant cases involving the replacement of hard tissues, several mechanical attributes, such as tensile strength, fatigue life, wear, Young's modulus, and ductility, are required (Xiao et al. 2017). In reality, many researchers have enhanced the mechanical properties of biomaterials for implant purposes. Wang et al. (2021) investigated the impact of surface coating on mechanical qualities in biomedical properties, finding that samples with coatings had better fracture toughness, hardness, and plasticity than samples that were not coated. Xiao et al. (2017) prepared coatings AlTiN via arc ion plating using the arc splitting technique.

5.5.3 CORROSION RESISTANCE

In biomedical applications, corrosion resistance is regarded as an important surface propriety that a material should possess in severe environments (Balan 2018).

Corrosion will have a significant impact on the implant's life span and performance. It is an important quality for dental materials, particularly because it prevents the surface from becoming rougher as well as affecting the aesthetics of the device (Sudha 2018). Uniform, galvanic, intergranular, and pitting corrosion are some of the several forms of corrosion. The liberation of metallic elements by corrosion will cause discoloration of adjacent soft tissue as well as a variety of allergic reactions. Many new titanium alloys have lately been created for use in medical implants. As a result, it's critical to look at the impacts of alloying elements on corrosion resistance. Songur et al. (2019) examined the corrosion behavior of Ti–29Nb–13Ta–4.6Zr in the laboratory, finding that oxide layer development was only submissive early in the corrosive process. On the surface, a bi-layer oxide coating was discovered, which increased osseointegration. More favorable effects against corrosion were observed due to the presence of Nb and Zr in larger quantities. The Ta and Nb-based alloys-8Ta-3Nb and Ti-10Ta-10Nb showed a comparable effect. Both alloys had excellent corrosion resistance; however, Ti-8Ta-3Nb had significantly poorer corrosion resistance than Ti-10Ta-10Nb. As a result of increasing the amount of alloying elements in Ti, successful suppression of oxygen growth occurred, resulting in a larger increase in corrosion resistance. Aside from Nb and Zr additions, additional metals such as Mo, Al, and Ta were also used to improve the material's stability for biomedical applications. The inclusion of Mo enhanced the stability of the passive film and significantly increased the alloy's passive range. The production of insoluble Mo chlorides accounted for the increase in stability. The inclusion of Al reduced the corrosion resistance of the material (Zhan et al. 2020). Roknian et al. (2018) studied the effects of ZnO nanoparticles using plasma electrolytic oxide coatings on pure titanium substrate. Artificially simulated solutions that resemble biological fluids present in the body were employed to examine the corrosion resistance of a created alloy. Miotto et al. (2016) evaluated the effects of plasma treatment of Ti-35Nb-7Zr-5Ta (TNZT) on biocompatibility. Artificial saliva was used to simulate Ti-35Nb-7Zr-5Ta over 5 to 20 years. The entire surface topography was characterized using SEM. The roughness of Ti-35Nb-7Zr-5Ta is unaffected by sodium fluoride, according to empirical findings. Simulated over 20 years in NaF, the Ti-6Al-4V alloy displayed an increase in roughness value, with scaling on its surface.

5.5.4 Wear Resistance

Friction, wear, and lubrication experimental research is becoming increasingly important in bioengineering. The corrosion characteristics of the material are the sole accelerating element in wear behavior. Researchers used friction wear and tensile tests to investigate the wear characteristics of newly produced Ti alloys. Bao et al. (2018) conducted a thorough investigation on the mechanical and physical properties of Ti alloy, as well as the challenges associated with long-term implantation. The abrasive wear behavior of the Ti implant was examined by Redmore et al. (2019). The surface hardness of the dental implant must be between 1000 and 1200 Vickers diamond pyramid hardness. Wang et al. (2021) conducted a wear study on CP Ti and Ti-6Al-7Nb against a Ti-6Al-4V ball using artificial saliva lubricated conditions. The imposed load, number of cycles, inclination angle, and material parameters all influence the wear behavior of Ti-6Al-7Nb alloy. Ti-6Al-7Nb outperformed pure Ti in terms of wear resistance. Adhesion wear and abrasive wear are two types of wear mechanisms for Ti alloy (Ureña et al. 2019). In comparison to air, SBF has a lower loss of Ti alloy owing to wear. As a result, human body wear is classified as an adhesive wear mechanism. The type of alloy and its microstructures determine the wear resistance. Surface hardening is the most effective way to increase wear resistance in Ti alloys. An examination was carried out on tissues that are closer to implanted failing hips. Ti-6Al-4V and polyethylene implants were used in this investigation. The wear on the surface of a femoral component made of Ti alloy was greater, resulting in implant loosening and the release of hazardous metal particles. Infection and implant loosening are greatly aided by metal debris. During the bone interaction, implants would experience cyclic micro-movements, resulting in considerable wear. Increased wear will result in the accumulation of harmful debris within the living organism (Gao et al. 2018).

Implant failure will occur as a result of wear debris loosening in a limited period of time. Each chewing action, including sliding motion in dental implants, is termed a tribo-corrosion cycle (Veerachamy et al. 2018). Chewing cycles erode away at the oxide coatings of Ti alloy, resulting in wear at the bone-implant interface. Wear will cause the implant to fail mechanically and cause undesirable biological effects. In a tribometer with an integrated optical profiler, reciprocating sliding corrosion studies were carried out. To slow down the rate of wear, several treatment methods were investigated, including heat treatment, surface coatings, and so on.

Ti-6Al-4V was given a heat treatment to test its wear resistance. The creation of a protective oxide layer during heat treatment resulted in a reduced wear rate in quenched Ti-6Al-4V alloy. Increasing the surface hardness was another way to improve wear resistance. As the hardness of the material increased, the contact area shrank. As a result, the percentage of adhesive wear dropped. Surface treatments influence the relationship between hardness and wear rate. Al_2O_3-coated samples have lower wear rates than alternative surface treatments. The higher wear resistance of Al_2O_3 coatings can be attributed to their superior coating thickness (Danışman et al. 2018).

5.5.5 Fatigue Characteristics

Due to the interdependence of different factors, determining a material's fatigue resistance is a difficult undertaking. The material's fatigue qualities are mostly determined by manufacturing process and alloying element (Fintová et al. 2021). The fatigue strength of implants is a major determinant of their longevity. According to Sazesh et al. (2019), testing an implant under stress conditions and simulated insertion is a time-consuming and costly operation. Dallago et al. (2018) looked at the fatigue properties of pure Ti dental implants. The effects of stress distribution on the surface treatment and exterior thread were studied. The fatigue test is performed for Ti fatigue curves acquired experimentally for treated and untreated surface. Surface treatment reduces sample fatigue strength, according to experimental results. Material quality, surface contact circumstances, CoF, and applied load all have an impact on the fretting fatigue strength. A fractographical analysis is performed for failure of materials (fatigue). Finally, investigations revealed that imperfections near the surface had an impact on fatigue behavior. Fatigue testing revealed a number of cracked locations. Although numerous surface modification techniques are employed to improve osseointegration, the impact of these treatments on implant fatigue is not taken into account. As a result, the implant's exterior condition has a significant impact on fatigue fracture nucleation (Sedmak et al. 2019).

5.5.6 OSSEOINTEGRATION

Osseointegration is defined as the formation of bone tissue surrounding an implant without the formation of fibrous tissue at the implant-bone interface (Apostu et al. 2018). It is, more precisely, the ability to grow normal tissue on a fixed implant in the host. However, fibril tissues form when the surface layer fails to incorporate with the bone, most commonly as a result of micro-motion that loosens the implant. Surface roughness, chemistry, and topography are all factors that have been linked to osseointegration (He et al. 2019). Furthermore, materials that are both bioactive and bio-inert are needed to improve implant osseointegration. Apart from that, surface modification techniques such as chemical etching, thermal spray, and blasting have been widely used to improve dental implant osseointegration (Wang et al. 2019).

5.6 COATING MATERIALS

5.6.1 TIN

Coating materials are applied either in liquid or solid form to the metal surfaces (as powders, laminates, or hot extrusions). This may be achieved before the process of metal forming, that is, as a coil or cut sheet or when the jar or end has a three-dimensional shape after forming. The mechanical properties of biomaterial coatings and the substrate (implant) are heavily dependent on the phase of film microstructure and deposition as well as the effect of interfacial constraints. Wan et al. (2004) fabricated TiN and Ti–O/TiN films by using plasma immersion ion implantation and deposition and deposited on Ti-6Al-4V substrates. In comparison to uncoated Ti–6Al–4V, TiN and Ti-O/TiN films have shown to have superior wear resistance. TiN coating material is a hard ceramic material, and it is a biocompatible material. This material is coated with SS 316L to improve some properties of stainless steel (SS). SS is used in biomedical applications as an alternative material because its hardness is poor. Saravanan et al. (2015) performed an optimization on the wear behavior of TiN coated SS 316L against Ti-6Al-4V alloy. During the test, the thickness, crystallography, surface hardness, and topography of coated surface were observed. The wear test was also examined using a ball on the disc by varying load, sliding velocity, and distance values. TiN coating shows better resistance to wear and CoF. The optimal parameters obtained to get minimum wear and CoF is 4N applied load, sliding speed of 0.5 m/s, and 1000 m distance for TiN-coated SS 316L materials Saravanan et al. (2015).

5.6.2 TIALN/ALTIN

Titanium aluminum nitride (metastatic thin hard coating materials containing more than 50% aluminum content) consists of nitrogen and aluminum and titanium metallic components. Titanium alloys have, sadly, insufficient abrasion resistance. The deposition of TiAlN or AlTiN on titanium-based implants for biomedical use was investigated by Walczak et al. (2019). Both TiAlN and AlTiN films were found to have good adhesion to a DMLS titanium-based substrate, as well as adequate morphological and mechanical properties.

5.6.3 ZrN

Zirconium nitride (ZrN) has excellent biocompatibility properties, corrosion resistance, good lubricity, hardness, and ductility. The combination of these multilayers will be a coating with good qualities such as a low friction, low wear rate, specific chemical protection, and strong corrosion resistance with good biocompatibility. Various processes, including cathodic arc evaporation and magnetron sputtering, can be used to create zirconium nitride coatings.

Its mechanical and chemical properties are affected by the composition of the coating. The surface roughness becomes a significant parameter for ZrN coatings applied as a biomaterial or an optical substance. A ZrN coating for an optical substance, surface roughness becomes an important characteristic. The coating properties could rely on layer thickness, elements, and nitrogen pressure.

5.6.4 DLC

DLC is also an important hard coating material mostly used in biomedical applications to improve surface properties without changing the bulk properties of the substrate material. There is a controversial issue on the performance of the DLC coating when applied to dental implantation. Antonowicz et al. (2020) investigated the mechanical properties of DLC-coated abutment screws made from Ti-6Al-4V alloy based on two hypothetical issues; the first is that the mechanical properties of coated and uncoated titanium alloy screws can be improved correspondingly using DLC coating in both nanoindentation testing and in-silico analysis and another hypothetical issue the uncoated cemented crowns are subjected to fatigue than the implant abutments coated with DLC. DLC is applied by using the radiofrequency plasma-activated (RFPA) method and unbalanced magnetron sputtering (UMS) method. The results show that two methods of coating techniques increase the Young's modulus and hardness of the abutment screw. The two hypothetical issues mean there is no effect of DLC coating on the abutment, and they concluded that the failure of the screw is due to the DLC coating method or its absence Antonowicz et al. (2020). In general, less cytotoxicity, good biocompatibility, and good mechanical properties have been observed, and DLC coatings are capable of improving chemical and mechanical properties (De Moura Silva et al. 2019).

An experimental investigation was done to propose the best application of steel materials coated by TiN and DLC. The was meant to contrast the wear and friction corrosion behavior of the TiN and DLC deposited by plasma-assisted methods to the substrate material (AISI 420 stainless steel) using X-ray diffraction and optical and electronic microscopy to characterize the microstructure of the coatings, a nanoindenter to measure Young's modulus and hardness, abrasive and wear tests performed by pin-on-disk, and the salt spray fog and electrochemical test in chloride solution used to evaluate the corrosion behavior. The test results shows that for sliding wear DLC is the best resistant coating material, while for abrasive wear, TiN is a good resistant material. According to the salt spray fogcorrosion test, both TiN and DLC show good corrosion resistance, whereas in the electrochemical test, TiN is less resistant compared to DLC. Thus, in the end, they concluded that DLC is the best coating material for AISI 420 stainless steel for different applications (Dalibón et al. 2019).

5.6.5 HYDROXYAPATITE

However, inflammatory reactions have been seen surrounding these implants as a result of the formation of an avascular fibrous tissue that encased them. To aid osseo-integration of these implants with surrounding tissues, a hydroxyapatite layer coating can be put on metal alloys. Due to its superior osseointegration and bioactive characteristics, hydroxyapatite is a desirable biomaterial for both dentistry and orthopedic applications (Bose et al. 2015).

5.7 TRIBOLOGICAL BEHAVIOR OF COATING MATERIALS

Wang et al. studied the tribological behavior of hard films CNx, DLC, and TiN deposited on NiTi alloys by PVD. Hard films, such as CNx, DLC, and TiN, were deposited on NiTi alloys to evaluate and optimize the hard film that may be used as surface protection for the NiTi alloy while maintaining adequate tribological properties. The scratch test and friction wear tester are used to systematically investigate the tribological behavior of hard thin films. It was shown that after the deposition of gradient films, the tribological efficiency of the NiTi alloy substratum was effectively improved. The Ti transition layer deposition helped to increase the strength of adhesion between the substrate and inorganic hard films, particularly for the CNx film. The Ti/CNx film had the best tribological efficiency (0.092 CoF) smooth surface test.

   The mechanical, chemical, and tribological characteristics of TiN coatings on Ti substrates for implant applications were studied by Uddin et al. (2019). Robust TiN coatings with 0.4μm average surface roughness showed the best adhesion, hardness, friction coefficient, corrosion, and wear rate in simulating bodily fluids with a 3.3-μm coating thickness. The higher substrate surface at 0.4 micrometer a 35% rise in hardness, 34% rise in adhesion, 23% reduction in COF, and 22% reduction wear rate compared to 0.1 micrometer substrate roughness. The expense of polishing an implant surface could be reduced by a coating of TiN with a thickness of 3–4 μm. TiN-coated implants with increased substrate surface roughness and a coating thickness of 3 to 4 μm could be employed to lower the cost of polishing an implant's surface.Dalibón et al. (2019) investigated the coating of DLC with plasma-assisted CVD and the deposition of TiN coatings by cathode arc PVD using martensitic stainless steel (AISI 420). The DLC exhibited superior sliding wear resistance, but the TiN exhibited superior abrasive wear resistance. The latter was associated with coating adherence or the resistance of the substrate–film interface to shear pressures. In terms of corrosion resistance, both coatings performed admirably in the salt spray test. The DLC coating performed better in electrochemical tests than the TiN coatings, which were probably not porous.

The wear properties of AlTiN-coated titanium alloys were investigated by Sivaprakasam et al. (2021). The input parameters were load, sliding speed, and sliding distance, whereas the response parameters were wear mass loss (WML) and CoF. The current work attempts to model WML and COF response empirically. WML and CoF were optimized using the desirability approach. Figure 5.9 shows the microscopy images of the wear tracks where abrasive wear is the primary mechanism, followed by adhesive wear. The study found that AlTiN can be used to cover titanium implants for biological applications. An SEM image of an AlTiN coating of titanium alloy is shown in Figure 5.10. A similar observation was also found for coating of AlTiN on316LVM stainless steel (Sivaprakasam et al. 2020).

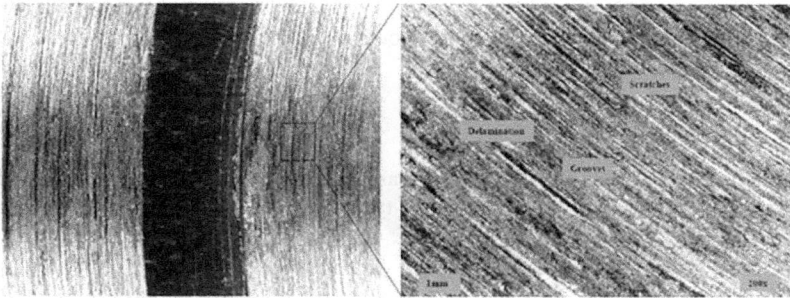

FIGURE 5.9   Microscopy images of wear tracks (Sivaprakasam et al. 2021).

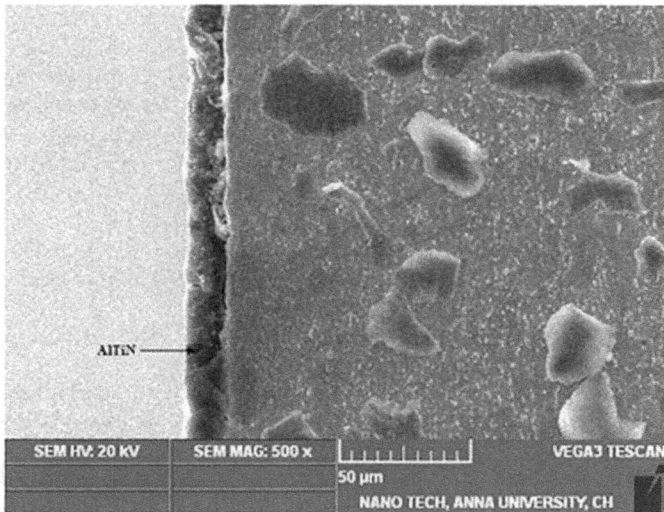

FIGURE 5.10   SEM image of AlTiN coating.

5.8   CONCLUDING REMARKS

Biomedical engineering is concerned with the use of implants and prostheses in the living human body. The usage of prostheses, particularly joint replacements, has increased globally as the complexity of human life has increased. As a result, a thorough examination of materials' tribological properties, such as friction and wear, is essential. PVD can be successfully applied for a variety of biomaterials for enhancement of tribological properties and biocompatibility.

This article discussed the most important characteristics of PVD for biomedical applications.

- The different types of coating process, bio materials, and coating materials were critically analyzed.
- Also, fundamental issues related to material properties influencing the implant surface were addressed.
- In addition to that, the tribological behavior of coating materials was also presented.

REFERENCES

Abdelrhman, Y., Gepreel, M. A. H., Kobayashi, S., Okano, S., & Okamoto, T. (2019). Biocompatibility of new low-cost (α+ β)-type Ti-Mo-Fe alloys for long-term implantation. *Materials Science and Engineering: C*, 99, 552–562.

Ankha, M. D. V. E. A., Silva, A. D. M., Prado, R. F. D., Camalionte, M. P., Vasconcellos, L. M. R. D., Radi, P. A., Silva, A. S. D., Vieira, L., & Carvalho, Y. R. (2019). Effect of DLC films with and without silver nanoparticles deposited on titanium alloy. *Brazilian Dental Journal*, 30, 607–616.

Antonowicz, M., Kurpanik, R., Walke, W., Basiaga, M., Sondor, J., & Paszenda, Z. (2020). Selected physicochemical properties of diamond like carbon (DLC) coating on Ti-13Nb-13Zr alloy used for blood contacting implants. *Materials*, 13(22), 5077.

Antunes, R. A., & De Oliveira, M. C. L. (2009). Corrosion processes of physical vapor deposition-coated metallic implants. *Critical Reviews in Biomedical Engineering*, 37(6).

Antunes, R. A., & De Oliveira, M. C. L. (2015). Effect of surface treatments on the fatigue life of magnesium and its alloys for biomedical applications. In *Surface Modification of Magnesium and Its Alloys for Biomedical Applications* (pp. 283–310). Woodhead Publishing.

Apostu, D., Lucaciu, O., Berce, C., Lucaciu, D., & Cosma, D. (2018). Current methods of preventing aseptic loosening and improving osseointegration of titanium implants in cementless total hip arthroplasty: A review. *Journal of International Medical Research*, 46(6), 2104–2119.

Balan, K. P. (2018). Chapter nine—Corrosion. In Balan, K. P. (Ed.), *Metallurgical Failure Analysis* (pp. 155–178). Elsevier.

Bao, M., Wang, X., Yang, L., Qin, G., & Zhang, E. (2018). Tribocorrosion behavior of Ti–Cu alloy in Hank's solution for biomedical application. *Journal of Bio-and Tribo-Corrosion*, 4(2), 1–4.

Behera, R. R., Das, A., Hasan, A., Pamu, D., Pandey, L. M., & Sankar, M. R. (2020). Effect of TiO_2 addition on adhesion and biological behavior of BCP-TiO_2 composite films deposited by magnetron sputtering. *Materials Science and Engineering: C*, 114, 111033. http://doi.org/10.1016/j.msec.2020.111033.

Bernhardt, A., Schneider, J., Schroeder, A., Papadopoulous, K., Lopez, E., Brückner, F., & Botzenhart, U. (2021). Surface conditioning of additively manufactured titanium implants and its influence on materials properties and in vitro biocompatibility. *Materials Science and Engineering: C*, 119, 111631.

Bose, S., Tarafder, S., & Bandyopadhyay, A. (2015). Hydroxyapatite coatings for metallic implants. In *Hydroxyapatite (Hap) for Biomedical Applications* (pp. 143–157). Woodhead Publishing.

Chen, Q., & Thouas, G. A. (2015). Metallic implant biomaterials. *Materials Science and Engineering R: Reports*, 87, 1–57.

Choi, A. H., Ben-Nissan, B., Bendavid, A., & Latella, B. (2016). Mechanical behavior and properties of thin films for biomedical applications. In *Thin Film Coatings for Biomaterials and Biomedical Applications* (pp. 117–141). Woodhead Publishing.

Dalibón, E. L., Pecina, J. N., Moscatelli, M. N., Ramos, M. A. R., Trava-Airoldi, V. J., & Brühl, S. P. (2019). Mechanical and corrosion behaviour of DLC and TiN coatings deposited on martensitic stainless steel. *Journal of Bio-and Tribo-Corrosion*, 5(2), 1–9.

Dallago, M., Fontanari, V., Torresani, E., Leoni, M., Pederzolli, C., Potrich, C., & Benedetti, M. (2018). Fatigue and biological properties of Ti-6Al-4V ELI cellular structures with variously arranged cubic cells made by selective laser melting. *Journal of the Mechanical Behavior of Biomedical Materials*, 78, 381–394.

Danışman, Ş., Odabas, D., & Teber, M. (2018). The effect of coatings on the wear behavior of Ti6Al4V alloy used in biomedical applications. In *IOP Conference Series: Materials Science and Engineering* (Vol. 295, No. 1, p. 012044). IOP Publishing.

De Moura Silva, A., de Figueiredo, V. M. G., do Prado, R. F., de Fátima Santanta-Melo, G., Ankha, M. D. V. E. A., de Vasconcellos, L. M. R., da Silva Sobrinho, A. S., Borges, A. L. S., & Junior, L. N. (2019). Diamond-like carbon films over reconstructive TMJ prosthetic materials: Effects in the cytotoxicity, chemical and mechanical properties. *Journal of Oral Biology and Craniofacial Research*, 9(3), 201–207.

Ding, H. H., Fridrici, V., Geringer, J., Fontaine, J., & Kapsa, P. (2019). Low-friction study between diamond-like carbon coating and Ti6Al4V under fretting conditions. *Tribology International*, 135, 368–388.

Dobrzański, L. A., Tański, T., Dobrzańska-Danikiewicz, A. D., Jonda, E., Bonek, M., & Drygała, A. (2015). Structures, properties and development trends of laser-surface-treated hot-work steels, light metal alloys and polycrystalline silicon. In *Laser Surface Engineering* (pp. 3–32). Woodhead Publishing.

Dong, H. (2003). Surface engineering in sport. In *Materials in Sports Equipment* (pp. 160–195). Woodhead Publishing.

Dong, M., Zhu, Y., Wang, C., Shan, L., & Li, J. (2019). Structure and tribocorrosion properties of duplex treatment coatings of TiSiCN/nitride on Ti6Al4V alloy. *Ceramics International*, 45(9), 12461–12468.

Du, J., Li, X., Li, K., Gu, X., Qi, W., and Zhang, K. (2016). High hydrophilic Si-doped TiO_2 nanowires by chemical vapor deposition. *Journal of Alloys Compounds*, 687, 893–897. http://doi.org/10.1016/j.jallcom.2016.06.182

El Hadad, A. A., García-Galván, F. R., Mezour, M. A., Hickman, G. J., Soliman, I. E., Jiménez-Morales, A., et al. (2020). Organic-inorganic hybrid coatings containing phosphorus precursors prepared by sol-gel on Ti6Al4V alloy: Electrochemical and in-vitro biocompatibility evaluation. *Progress in Organic Coatings*, 148, 105834. http://doi.org/10.1016/j.porgcoat.2020.105834

Fintová, S., Dlhý, P., Mertová, K., Chlup, Z., Duchek, M., Procházka, R., & Hutař, P. (2021). Fatigue properties of UFG Ti grade 2 dental implant vs. conventionally tested smooth specimens. *Journal of the Mechanical Behavior of Biomedical Materials*, 123, 104715.

Gangatharan, K., Selvakumar, N., Narayanasamy, P., & Bhavesh, G. (2016). Mechanical analysis and high temperature wear behaviour of AlCrN/DLC coated titanium alloy. *International Journal of Surface Science and Engineering*, 10(1), 27–40.

Gao, A., Hang, R., Bai, L., Tang, B., & Chu, P. K. (2018). Electrochemical surface engineering of titanium-based alloys for biomedical application. *Electrochimica Acta*, 271, 699–718.

Gepreel, M. A. H., & Niinomi, M. (2013). Biocompatibility of Ti-alloys for long-term implantation. *Journal of the Mechanical Behavior of Biomedical Materials*, 20, 407–415.

Gnanavel, S., Ponnusamy, S., & Mohan, L. (2018). Biocompatible response of hydroxyapatite coated on near-β titanium alloys by E-beam evaporation method. *Biocatalysis and Agricultural Biotechnology*, 15, 364–369.

Gubbi, P., & Wojtisek, T. (2018). The role of titanium in implant dentistry. In *Titanium in Medical and Dental Applications* (pp. 505–529). Woodhead Publishing.

He, W., Yin, X., Xie, L., Liu, Z., Li, J., Zou, S., & Chen, J. (2019). Enhancing osseointegration of titanium implants through large-grit sandblasting combined with micro-arc oxidation surface modification. *Journal of Materials Science: Materials in Medicine*, 30(6), 1–11.

Im, G. I. (2020). Biomaterials in orthopaedics: The past and future with immune modulation. *Biomaterials Research*, 24(1), 1–4.

Jammalamadaka, U., & Tappa, K. (2018). Recent advances in biomaterials for 3D printing and tissue engineering. *Journal of Functional Biomaterials*, 9(1), 22.

Jayaraj, K., & Pius, A. (2018). Biocompatible coatings for metallic biomaterials. In *Fundamental Biomaterials: Metals* (pp. 323–354). Woodhead Publishing.

Ji, H., Xia, L., Ma, X., Sun, Y., & Sun, M. (1999). Tribological behaviour of duplex treated Ti–6Al–4V: Combining nitrogen PSII with a DLC coating. *Tribology International*, 32(5), 265–273.

Jin, J., Li, X. H., Wu, J. W., & Lou, B. Y. (2018). Improving tribological and corrosion resistance of Ti6Al4V alloy by hybrid microarc oxidation/enameling treatments. *Rare Metals*, 37(1), 26–34.

Kaur, M., & Singh, K. (2019). Review on titanium and titanium based alloys as biomaterials for orthopaedic applications. *Materials Science and Engineering: C*, 102, 844–862.

Kheradmandfard, M., Kashani-Bozorg, S. F., Lee, J. S., Kim, C. L., Hanzaki, A. Z., Pyun, Y. S., Cho, S. W., Amanov, A., & Kim, D. E. (2018). Significant improvement in cell adhesion and wear resistance of biomedical β-type titanium alloy through ultrasonic nanocrystal surface modification. *Journal of Alloys and Compounds*, 762, 941–949.

Khorasani, A. M., Goldberg, M., Doeven, E. H., & Littlefair, G. (2015). Titanium in biomedical applications—Properties and fabrication: A review. *Journal of Biomaterials and Tissue Engineering*, 5(8), 593–619.

Kopova, I., Kronek, J., Bacakova, L., & Fencl, J. (2019). A cytotoxicity and wear analysis of trapeziometacarpal total joint replacement implant consisting of DLC-coated Co-Cr-Mo alloy with the use of titanium gradient interlayer. *Diamond and Related Materials*, 97, 107456.

Li, Junlei, Qin, Ling, Yang, Ke, Ma, Zhijie, Wang, Yongxuan, Cheng, Liangliang, Zhao, Dewei (2020). Materials evolution of bone plates for internal fixation of bone fractures: A review. *Journal of Materials Science & Technology*, 36, 190–208.

Liang, S. X., Yin, L. X., Zheng, L. Y., Xie, H. L., Yao, J. X., Ma, M. Z., & Liu, R. P. (2018). Tribological behavior and wear mechanism of TZ20 titanium alloy after various treatments. *Journal of Materials Engineering and Performance*, 27(9), 4645–4654.

Lin, X., et al. (2018). Biocompatibility of bespoke 3D-printed titanium alloy plates for treating acetabular fractures. *BioMed Research International*, 2018, 2053486.

Liu, D., Yang, T., Ma, H., & Liang, Y. (2020). The microstructure, bio-tribological properties, and biocompatibility of titanium surfaces with graded zirconium incorporation in amorphous carbon bioceramic composite films. *Surface and Coatings Technology*, 385, 125391. http://doi.org/10.1016/j.surfcoat.2020.125391

Liu, X., Chu, P. K., & Ding, C. (2004). Surface modification of titanium, titanium alloys, and related materials for biomedical applications. *Materials Science and Engineering: R: Reports*, 47(3–4), 49–121.

Liu, Y., Bian, D., Zhao, Y., & Wang, Y. (2019). Influence of curing temperature on corrosion protection property of chemically bonded phosphate ceramic coatings with nano-titanium dioxide reinforcement. *Ceramics International*, 45(2), 1595–1604.

Ma, W., Ruys, A. J., & Zreiqat, H. (2009). Diamond-like carbon (DLC) as a biocompatible coating in orthopaedic and cardiac medicine. *Cellular Response to Biomaterials*, 391–426.

Martín-Palma, Raúl J., and Lakhtakia, Akhlesh (2013). *Engineered Biomimicry: Chapter 15—Vapor-Deposition Techniques*. Elsevier.

Michelmore, A. (2016). Fundamentals of thin film technologies for biomedical applications. In *Thin Film Growth on Biomaterial Surfaces*. Elsevier Ltd.

Ming'e, W., Guojia, M., Xing, L., & Chuang, D. (2016). Morphology and mechanical properties of TiN coatings prepared with different PVD methods. *Rare Metal Materials and Engineering*, 45(12), 3080–3084.

Miotto, L. N., Fais, L. M., Ribeiro, A. L., & Vaz, L. G. (2016). Surface properties of Ti-35Nb-7Zr-5Ta: Effects of long-term immersion in artificial saliva and fluoride solution. *The Journal of Prosthetic Dentistry*, 116(1), 102–111.

Muthuraja, A., Naik, S., Rajak, D. K., & Pruncu, C. I. (2019). Experimental investigation on chromium-diamond like carbon (Cr-DLC) coating through plasma enhanced chemical vapour deposition (PECVD) on the nozzle needle surface. *Diamond and Related Materials*, 100, 107588.

Nasker, P., & Sinha, A. (2018). Titanium based bulk metallic glasses for biomedical applications. In *Fundamental Biomaterials: Metals* (pp. 269–283). Woodhead Publishing.

Nouri, A., & Wen, C. (2015). Introduction to surface coating and modification for metallic biomaterials. *Surface Coating and Modification of Metallic Biomaterials*, 3–60.

Oliveira, L. Y., Kuromoto, N. K., & Siqueira, C. J. (2014). Treating orthopedic prosthesis with diamond-like carbon: Minimizing debris in Ti6Al4V. *Journal of Materials Science: Materials in Medicine*, 25(10), 2347–2355.

Polyakova, V. V., Semenova, I. P., Polyakov, A. V., Magomedova, D. K., Huang, Y., & Langdon, T. G. (2017). Influence of grain boundary misorientations on the mechanical behavior of a near-α Ti-6Al-7Nb alloy processed by ECAP. *Materials Letters*, 190, 256–259.

Ratner, B. D. (2019). Biomaterials: Been there, done that, and evolving into the future. *Annual Review of Biomedical Engineering*, 21, 171–191.

Raval, N., Kalyane, D., Maheshwari, R., & Tekade, R. K. (2019). Surface modifications of biomaterials and their implication on biocompatibility. In *Biomaterials and Bionanotechnology* (pp. 639–674). Academic Press.

Redmore, E., Li, X., & Dong, H. (2019). Tribological performance of surface engineered low-cost beta titanium alloy. *Wear*, 426, 952–960.

Roknian, M., Fattah-Alhosseini, A., Gashti, S. O., & Keshavarz, M. K. (2018). Study of the effect of ZnO nanoparticles addition to PEO coatings on pure titanium substrate: microstructural analysis, antibacterial effect and corrosion behavior of coatings in Ringer's physiological solution. *Journal of Alloys and Compounds*, 740, 330–345.

Roy, R. K., & Lee, K. R. (2007). Biomedical applications of diamond-like carbon coatings: A review. *Journal of Biomedical Materials Research Part B: Applied Biomaterials*, 83(1), 72–84.

Saba, F., Zhang, F., Liu, S., & Liu, T. (2018). Tribological properties, thermal conductivity and corrosion resistance of titanium/nanodiamond nanocomposites. *Composites Communications*, 10, 57–63.

Sandlöbes, S., Korte-Kerzel, S., & Raabe, D. (2019). On the influence of the heat treatment on microstructure formation and mechanical properties of near-α Ti-Fe alloys. *Materials Science and Engineering: A*, 748, 301–312.

Saravanan, I., Perumal, A. E., Vettivel, S. C., Selvakumar, N., & Baradeswaran, A. (2015). Optimizing wear behavior of TiN coated SS 316L against Ti alloy using response surface methodology. *Materials & Design*, 67, 469–482.

Sazesh, S., Ghassemi, A., Ebrahimi, R., & Khodaei, M. (2019). Fabrication and characterization of nHA/titanium dental implant. *Materials Research Express*, 6(4), 045060.

Sedmak, A., Čolić, K., Grbović, A., Balać, I., & Burzić, M. (2019). Numerical analysis of fatigue crack growth of hip implant. *Engineering Fracture Mechanics*, 216, 106492.

Shanaghi, A., and Chu, P. K. (2019b). Investigation of corrosion mechanism of NiTi modified by carbon plasma immersion ion implantation (C-PIII) by electrochemical impedance spectroscopy. *Journal of Alloys Compounds*, 790, 1067–1075. http://doi.org/10.1016/j.jallcom.2019.03.272

Shim, J. W., Bae, I. H., Jeong, M. H., Park, D. S., Lim, K. S., Kim, J. U., Kim, M. K., Kim, J. H., Kim, J. H., & Sim, D. S. (2019). Effects of a titanium dioxide thin film for improving the biocompatibility of diamond-like coated coronary stents. *Metals and Materials International*, 1–8.

Sivaprakasam, P., Elias, G., Prabu, P. M., & Balasubramani, P. (2020). Experimental investigations on wear properties of AlTiN coated 316LVM stainless steel. *Materials Today: Proceedings*, 33, 3470–3474.

Sivaprakasam, P., Kirubel, A., Elias, G., Maheandera Prabu, P., & Balasubramani, P. (2021). Mathematical modeling and analysis of wear behavior of AlTiN coating on titanium alloy (Ti-6Al-4V). *Advances in Materials Science and Engineering*, 2021.

Sommer, U., Laurich, S., de Azevedo, L., Viehoff, K., Wenisch, S., Thormann, U., & Schnettler, R. (2020). In Vitro and in Vivo biocompatibility studies of a cast and coated titanium alloy. *Molecules*, 25(15), 3399.

Songur, F., Dikici, B., Niinomi, M., & Arslan, E. (2019). The plasma electrolytic oxidation (PEO) coatings to enhance in-vitro corrosion resistance of Ti–29Nb–13Ta–4.6 Zr alloys: The combined effect of duty cycle and the deposition frequency. *Surface and Coatings Technology*, 374, 345–354.

Sudha, P. N., Sangeetha, K., Jisha Kumari, A. V., Vanisri, N., & Rani, K. (2018). Corrosion of ceramic materials. In Thomas, S., Balakrishnan, P., and Sreekala, M.S. (Eds.), *Fundamental Biomaterials: Ceramics* (pp. 223–250). Woodhead Publishing.

Tanzi, M. C., Farè, S., & Candiani, G. (2019). Biomaterials and applications. In *Foundations of Biomaterials Engineering* (pp. 199–287). Academic Press.

Thorwarth, G., Falub, C. V., Müller, U., Weisse, B., Voisard, C., Tobler, M., & Hauert, R. (2010). Tribological behavior of DLC-coated articulating joint implants. *Acta Biomaterialia*, 6(6), 2335–2341.

Tüten, N., Canadinc, D., Motallebzadeh, A., & Bal, B. U. R. A. K. (2019). Microstructure and tribological properties of TiTaHfNbZr high entropy alloy coatings deposited on Ti6Al4V substrates. *Intermetallics*, 105, 99–106.

Uddin, G. M., Jawad, M., Ghufran, M., Saleem, M. W., Raza, M. A., Rehman, Z. U., Arafat, S. M., Irfan, M., & Waseem, B. (2019). Experimental investigation of tribo-mechanical and chemical properties of TiN PVD coating on titanium substrate for biomedical implants manufacturing. *The International Journal of Advanced Manufacturing Technology*, 102(5), 1391–1404.

Ureña, J., Tabares, E., Tsipas, S., Jiménez-Morales, A., & Gordo, E. (2019). Dry sliding wear behaviour of β-type Ti-Nb and Ti-Mo surfaces designed by diffusion treatments for biomedical applications. *Journal of the Mechanical Behavior of Biomedical Materials*, 91, 335–344.

Veerachamy, S., Hameed, P., Sen, D., Dash, S., & Manivasagam, G. (2018). Studies on mechanical, biocompatibility and antibacterial activity of plasma sprayed nano/micron ceramic bilayered coatings on Ti–6Al–4V alloy for biomedical application. *Journal of Nanoscience and Nanotechnology*, 18(7), 4515–4523.

Wadood, A. (2016). Brief overview on nitinol as biomaterial. *Advances in Materials Science and Engineering*, 2016.

Walczak, M., Pasierbiewicz, K., & Szala, M. (2019). Adhesion and mechanical properties of TiAlN and AlTiN magnetron sputtered coatings deposited on the DMSL titanium alloy substrate. *Acta Physica Polonica, A*, 136(2).

Wan, G. J., Huang, N., Leng, Y. X., Yang, P., Chen, J. Y., Wang, J., & Sun, H. (2004). TiN and Ti–O/TiN films fabricated by PIII-D for enhancement of corrosion and wear resistance of Ti–6Al–4V. *Surface and Coatings Technology*, 186(1–2), 136–140.

Wang, C., Hu, H., Li, Z., Shen, Y., Xu, Y., Zhang, G., Zeng, X., Deng, J., Zhao, S., Ren, T., & Zhang, Y. (2019). Enhanced osseointegration of titanium alloy implants with laser microgrooved surfaces and graphene oxide coating. *ACS Applied Materials & Interfaces*, 11(43), 39470–39483.

Wang, G., Wang, S., Yang, X., Yu, X., Wen, D., Chang, Z., & Zhang, M. (2021). Fretting wear and mechanical properties of surface-nanostructural titanium alloy bone plate. *Surface and Coatings Technology*, 405, 126512.

Wilson, J. (2018). Metallic biomaterials: State of the art and new challenges. *Fundamental Biomaterials: Metals*, 1–33.

Xiao, B. J., Chen, Y., Dai, W., Kwork, K. Y., Zhang, T. F., Wang, Q. M., Wang, C. Y., & Kim, K. H. (2017). Microstructure, mechanical properties and cutting performance of AlTiN coatings prepared via arc ion plating using the arc splitting technique. *Surface and Coatings Technology*, 311, 98–103.

Yang, L. (2017). Nanotechnology-enhanced orthopedic materials: Fabrications, applications and future trends. *Mrs Bulletin*, 42.

Yang, P., Huang, N., Leng, Y., Wan, G., Zhao, A., Chen, J., et al. (2007). Functional inorganic films fabricated by PIII(-D) for surface modification of blood contacting biomaterials: Fabrication parameters, characteristics and antithrombotic properties. *Surface and Coatings Technology*, 201, 6828–6832. http://doi.org/10.1016/j.surfcoat.2006.09.014.

Youn, Y. H., Lee, S. J., Choi, G. R., Lee, H. R., Lee, D., Heo, D. N., et al. (2019). Simple and facile preparation of recombinant human bone morphogenetic protein-2 immobilized titanium implant via initiated chemical vapor deposition technique to promote osteogenesis for bone tissue engineering application. *Materials Science & Engineering C-Materials for Biological Applications*, 100, 949–958. http://doi.org/10.1016/j.msec.2019.03.048.

Yu, L., Jin, G., Ouyang, L., Wang, D., Qiao, Y., & Liu, X. (2016). Antibacterial activity, osteogenic and angiogenic behaviors of copper-bearing titanium synthesized by PIII&D. *Journal of Materials Chemistry B*, 4, 1296–1309. http://doi.org/10.1039/c5tb02300a.

Zhan, W., Qian, X., Gui, B., Liu, L., Liu, X., Li, Z., & Hu, L. (2020). Preparation and corrosion resistance of titanium—zirconium—cerium based conversion coating on 6061 aluminum alloy. *Materials and Corrosion*, 71(3), 419–429.

Ziabka, M., Kiszka, J., Trenczek-Zajac, A., Radecka, M., Cholewa-Kowalska, K., Bissenik, I., et al. (2020). Antibacterial composite hybrid coatings of veterinary medical implants. *Materials Science & Engineering C-Materials for Biological Applications*, 112, 110968. http://doi.org/10.1016/j.msec.2020.110968.

6 Friction Stir Welding
A Sustainable Procedure for Joining Steels

Anmol Bhatia, Reeta Wattal

CONTENTS

6.1 INTRODUCTION

Friction stir welding (FSW) is a solid-state welding process that involves no melting of the workpiece. A non-consumable tool is used in the process that has a small pin projecting from the shoulder portion of the tool (Mishra & Ma, 2005). The plates that are to be welded are rigidly fixed using a fixture, and a backing plate is used to support the plates together. The rotating tool is plunged into the joint of the plates through the pin until the shoulder touches the surface of the workpiece. Because of this rotation of the tool, localized heat is produced that leads to material softening around the pin. The tool movements comprise two actions: rotation and translation. Because of the translatory motion, the material moves from the front to the back of the pin, thereby producing a solid-state joint. Then the tool

DOI: 10.1201/9781003269298-6

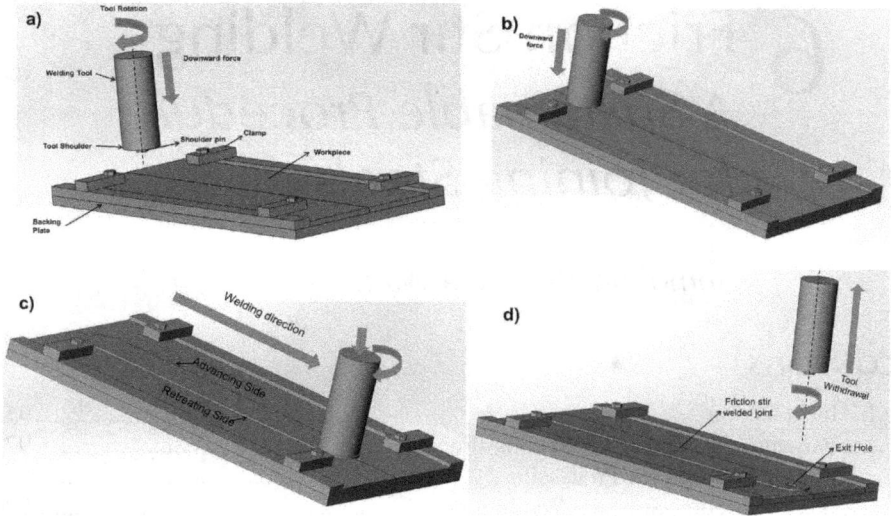

FIGURE 6.1 Working principle of friction stir welding.

is withdrawn, leaving the exit hole. Figure 6.1 illustrates the working principle of friction stir welding.

Joining of steels is mostly done using liquid state welding that incorporates defects such as hydrogen embrittlement, hot cracking, and so on. Compared to fusion welding, friction stir welding is capable of producing defect-free steel joints with improved microstructural and mechanical properties (Küçükömeroğlu et al., 2018). Apart from this, the friction stir process offers many advantages: (1) In comparison to the conventional welding processes, filler material is not required in the friction stir welding process. This filler material requirement does not even add to the cost, but it leads to a lot of energy wastage to melt the filler material. Thus, friction stir welding is sustainable in this aspect. (2) The friction stir welding process is an automated process in which the work is carried out by specialized equipment that involves less human intervention. This would lead to zero variation in the welded jobs. (3) Friction stir welding is a greener welding process, as it is energy efficient and does not require any filler material. It does not produce any toxic fumes, slag, or UV radiation. In addition to this, it also produces very little noise during its operation in comparison to conventional arc welding process. (4) Joints made by friction stir welding process have very good surface quality; hence, an extra effort of removal of reinforcements in conventional arc welding process is required in this process. (5) By using the proper equipment, the process can be easily used in difficult to weld positions like horizontal, vertical, or overhead positions. (6) Welds can be made at very high welding speed and with theoretically a 100% duty factor, as no human fatigue is involved. Since welds are usually made in a single run and joint preparation is not needed, productivity increases manyfold. The productivity of the process improves because no edge preparation is needed, and no subsequent finishing operations are usually required to be carried out.

Although friction stir welding is successfully implemented for welding aluminum, friction stir welding of steels has not reached the same level of commercialization as FSW of aluminum because of the following challenges: (1) selection of proper tool material for welding steel, (2) elimination of defects occurring during friction stir welding of steels due to high temperature and forces, (3) selecting the range of process parameters, (4) tool deterioration rate is high, and (5) optimization of the process parameters to produce defect-free friction stir-welded steel joints (Karami et al., 2021). In the last decade, the welding of steels using the friction stir technique has gained popularity among researchers. Looking at recent developments, it is critical to summarize the technological aspects of joining steels using FSW. The present chapter aims to showcase the challenges reported and solutions developed while joining steels using FSW and to demonstrate that it is a sustainable technique when applied to joining steel structures.

6.2 VARIANTS OF FRICTION STIR WELDING PROCESS

Over the last decade, there have been several variants of the friction stir welding process reported in the literature. The stirring of the work material in a solid state using a non-consumable tool is a common feature of all variants. Many advantages of the friction stir welding process are retained in these variants. Some of the variants are described in the following paragraphs.

6.2.1 FRICTION-STIR PROCESSING

With the advances of friction stir welding, a new process named friction stir processing (FSP) has also been developed in the recent past. It is based on the principle of friction stir welding in the sense that the material being processed is subjected to mechanical working and heating. In this process, a slightly modified tool is allowed to work on the material with the intention of changing its grain size and mechanical properties on its surface. This process is applicable in the production of composite materials.

The initial work on FSP was reported by Mishra et al. wherein they demonstrated the technique of fabrication of surface composites (Mishra et al., 2003). Since then the process has been used for fabricating materials like Mg-Al-Zn alloy (Feng & Ma, 2007), magnesium-based nano-composites (Sunil et al., 2016), nano-tool steel (Ghasemi-kahrizsangi et al., 2015), Al-Ni intermetallic composites (Ke et al., 2010), and particle-reinforced aluminum matrix composites (Ardalanniya et al., 2021). It has been used to enhance the mechanical properties in aluminum alloys (McNelley et al., 2008; Santella et al., 2005), ferrous alloys (Mironov et al., 2008), magnesium alloys (Wang, Han et al., 2020), and copper (Cartigueyen & Mahadevan, 2015).

6.2.2 FRICTION-STIR SPOT WELDING

Another variant which is suited for producing lap joints is friction stir spot welding (FSSW). In this process, a tool with pin length less than the combined thickness of plates in lap joint is plunged into the workpiece. The tool is kept rotating for some time at the same spot it was inserted until sufficient heat has been produced. Plastic flow of material takes place beneath the tool pin. All this time, the material is under strong compressive force from tool shoulder. After the tool is withdrawn, a strong solid-state bond is formed between the top and bottom plate. Figure 6.2 shows the friction stir spot welding stages (Lakshminarayanan et al., 2015). The paper presented a detailed investigation of friction stir-welded low-carbon automotive steel in which enhanced mechanical properties were reported.

A lot of work has also been reported on friction stir spot welding for aluminum alloys (Suryanarayanan & Sridhar, 2020). This process has been demonstrated to form micro-welds (Wang et al., 2010). Friction-stir spot welding of ferrous alloys (Miles et al., 2011), bulk metallic glass (Ji et al., 2009), and even fiber-reinforced polymers (Ji et al., 2009) has been studied.

6.2.3 Hybrid Friction Stir Welding

In this type of variant, an auxiliary heat source to assist the heating during friction stir welding is arranged (Sun et al., 2013). The addition of heat softens the material and thereby makes welding at higher speeds possible for strong materials. Also, the amount of force experienced during welding is reduced. The most common heat source can be a laser or a gas torch. Figure 6.3 shows a hybrid FSW machine with

Plunging   Stirring   Retracting

FIGURE 6.2   Different stages during FSSW of low carbon automotive steel (Lakshminarayanan et al., 2015).

Source: © [Elsevier]. Reproduced by permission of Elsevier

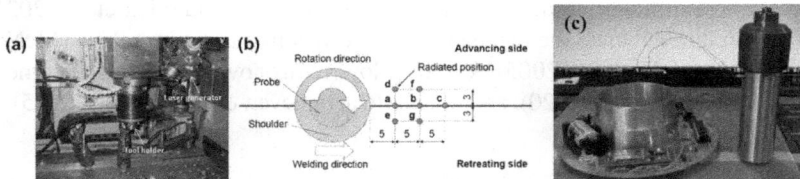

FIGURE 6.3   a) Photo of the FSW machine equipped with a laser preheating system; b) schematic diagram showing the various positions of focal points of the laser beam; c) photo of the temperature measurement device (Sun et al., 2013).

Source: © [Elsevier]. Reproduced by permission of Elsevier

arrangements to measure the temperature. Song et al. successfully welded Inconel 600 alloy at a welding speed 1.5 times faster than the maximum speed possible for friction stir welding alone. The paper reported a reduction in grain size during hybrid welding (Song et al., 2009).

6.2.4 FRICTION-STIR WELDING WITH BOBBIN TOOL

In this variant, a special tool is used that has shoulders on both sides (Sued et al., 2014). The bobbin tool is shown in Figure 6.4. The gap between the shoulders is kept equal to the thickness of the plates being welded. This process is very useful for welding thick sections. The problem of a keyhole occurring at the end of the weld is also eliminated.

Many other variants like friction-stir knead welding (Geiger et al., 2008), resistance friction stir welding (Luo et al., 2009), and so on are being studied regularly, which shows the state-of-the-art research and development activities going on in the area of friction stir welding.

(a)   (b)

FIGURE 6.4 Bobbin friction stir welding tools used during experiments: a) changeable pin; b) small tool. Both tools have scroll shoulders (Sued et al., 2014).

Source: © [Elsevier]. Reproduced by permission of Elsevier

6.3   TOOL MATERIALS

There are two concerns about steel: steel does not soften below 1000°C, and its yield strength can go up to 700 MPa. These two concerns need to be addressed properly when selecting the tool material. The important properties that need to be considered for selecting the tool are the strength of the tool, hardness of the tool, wear resistance, fatigue resistance, thermal conductivity, toughness, and chemical stability. The materials that have been tried by researchers for welding steel are tabulated in Table 6.1. The types of materials with satisfactory results for welding steel using the friction stir welding process are refractory metals and super-abrasive tools.

TABLE 6.1
Summary of Tool Materials Used for Friction Stir Welding of Steel

Welding Performed on Base Metal	Joint Type	Thickness	Tool Material	Reference
AISI 1018	Butt joint	0.25 inch	Tungsten-based alloys	(Lienert et al., 2003)
Low carbon steel	Butt joint	1.5 mm	Tungsten carbide	(Ueji et al., 2006)
DP590 steel	Butt joint	1.5 mm	PCBN	(Miles et al., 2006)
L80 steel	Lap joint	—	Tungsten rhenium alloy and potassium-doped pure alloy	(Gan et al., 2007)
AISI 1012 and AISI 1035	Butt joint	1.6 mm	Tungsten carbide	(Fujii et al., 2006)
RQT-701	Butt joint	12.5 mm	Tungsten-rhenium alloy	(Barnes et al., 2008)
High-strength steel and ultrahigh-strength steel	Butt joint	—	PCBN	(Miles et al., 2009)
M190 steel	Butt joint	1 mm	Hybrid carbide	(Ghosh et al., 2010)
304 stainless steels	Butt joint	6 mm	PCBN	(Park et al., 2003)
Super-austenitic stainless steel	Butt joint	6.35 mm	Tungsten carbide	(Sato et al., 2005)
Duplex stainless steel	Butt joint	4 mm	PCBN	(Klingensmith et al., 2005)
Oxide-dispersed strengthened (ODS) steel	Butt joint	—	PCBN	(Legendre et al., 2009)
AISI 304 austenitic stainless steel	Butt welded	2.95 mm	Tungsten carbide with cobalt	(Siddiquee & Pandey, 2014)
DH36 steel plates.	Butt welded	300 × 100 × 4 mm	WC-6 wt.% Co and WC-10 wt.% Co	(Tiwari et al., 2019)
SC45 steel	Butt welded	2.4 mm thick	WC-12Co	(Avettand-Fènoël et al., 2019)

6.3.1 Refractory Metals

This category of tool material includes tungsten carbide, tungsten carbide cobalt alloy, silicon nitride, and tungsten lanthanide (Muhammad et al., 2018). The properties of refractory tool materials are shown in Table 6.2.

Tungsten and molybdenum were used in the initial stages of friction stir welding because these materials have sufficient hot hardness property, good creep resistance, and high tensile strength, but the major limitation of these tools was tool fracture in the plunging stage (Lienert et al., 2003). Plunging at the start of welding at room temperature resulted in brittle fracture of the tool. This caused a need to preheat the workpiece material above ductile-brittle transition temperature before the initial plunge of the tool. Later, a tungsten-rhenium tool was used for friction stir welding of steel. The main benefit of the tool was that there was a reduced ductile to brittle transition temperature. Thus, pre-heating of the tool was not needed when a tungsten-rhenium alloy was used. Although the addition of rhenium leads to a decrease in wear resistance of the tool, tool wear was still a problem. This led to tungsten's inclusion in the weld, which would seriously affect the mechanical properties. The refractory materials used as tools currently are tungsten carbide and tungsten carbide with cobalt (Bhatia & Wattal, 2021). Tungsten carbide is one of the most suitable choices for researchers, as it can withstand a temperature of 2800 °C and is very hard and wear resistant. The issue with this material is its brittle nature, which makes it difficult to be used as such. Cobalt can be used in the composite to an extent of 5–25%. This composite of tungsten carbide with cobalt is also a popular choice of researchers, as shown in Table 6.1.

6.3.2 Synthetic Materials

Polycrystalline cubic boron nitride (PCBN) is the tool that is currently used for friction stir welding of steels. PCBN is one of the hardest materials, second to diamond in terms of hardness. This material can withstand very high temperatures. The properties of PCBN material are shown in Table 6.3. PCBN tool materials are created with high-pressure, high-temperature technology, bonding cubic boron nitride crystals together with metallic or ceramic binders forming a substrate, which will provide strength to the PCBN composite. PCBN is the most promising material for making friction stir welding tools. Numerous researchers have found PCBN satisfactory,

TABLE 6.2
Properties of Refractory Materials

Tool Material\ Property	Melting Point [°C]	Boiling Point [°C]	Young's Modulus [GPa]	Vicker's Hardness [MPa]
Tungsten	3422	5555	411	3430
Rhenium	3200	5560	460	2500
Molybdenum	2625	4640	330	1530

as shown in Table 6.1. The only problem with PCBN is tool breakage under non-uniform loads.

6.3.3 TOOL PROFILES

Tool pin profile is a very important element to produce a defect-free weld. This also governs the material flow during the friction stir welds. This tool pin is also responsible for producing heat. At the initial phase of welding, the tool would be plunged; the part that is responsible for providing the heat is the tool pin. It is also responsible for producing the shape and size of the weld nugget. The tool pin should be such that it stirs the material properly and generates the appropriate amount of heat. At the same time, the shape should be such that it is long lasting and penetrates the cold workpiece during the start of welding. Based on these considerations, pin profiles are of the following types: cylindrical (straight cylindrical, taper cylindrical, and threaded cylindrical), square, and triangular (Elangovan & Balasubramanian, 2008). Figure 6.5 shows different types of tool pin profiles used for friction stir welding research.

TABLE 6.3
Properties of PCBN Tools

Density (g/cc)	Fracture Toughness $\left[\text{MPa}\sqrt{m}\right]$	Thermal Conductivity [W/m °C]	Vicker's Hardness [HV]
3.4	7.9	120	10000

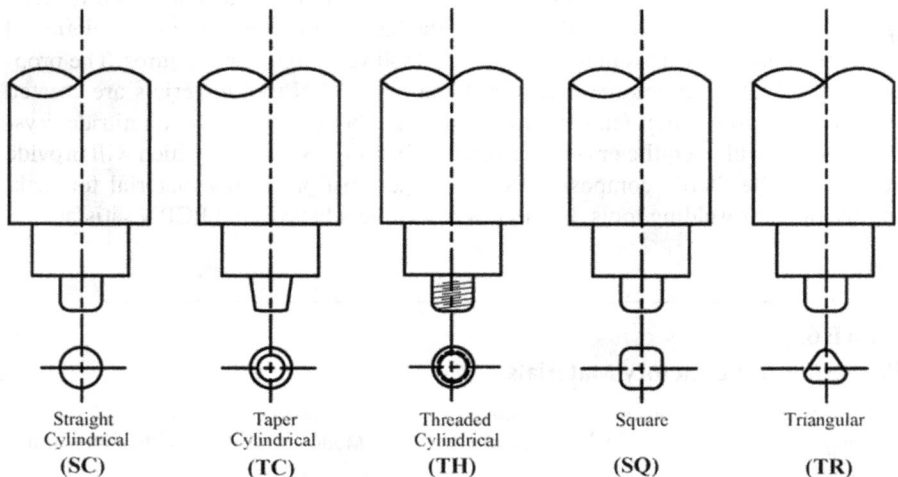

FIGURE 6.5 Various tool pin profiles (Elangovan & Balasubramanian, 2008).

Source: © [Elsevier]. Reproduced by permission of Elsevier

6.4 MICROSTRUCTURAL CHARACTERIZATION

Microstructural analysis is helpful in analyzing the weld structure and understanding the flow of the material. Microstructural analysis is also helpful in interpreting any defects occurring in the welded structures. The grain structure of the weld can be checked by comparing it with the parent metal. In recent practice, majority of research is based on improving the mechanical properties of the welds. These properties can be improved by analyzing the homogeneity of the grains (Küçükömeroğlu et al., 2018; Yabuuchi et al., 2014). The important process parameters that affect the microstructure of the friction stir–welded joints are welding speed, pin geometry, tool RPM, shoulder diameter, and tool tilt angle. Changing the process parameters would lead to generation of different amounts of heat. This generated heat will no doubt affect the microstructure (Avila et al., 2016; Cho et al., 2011). The main factor that influences the material flow is tool geometry. Shoulder diameter affects the uppermost layer of material flow, and the tool pin dominates the movement of materials around the middle and lower segment of weldment (Imam et al., 2016; Wang, Zhang et al., 2020)

AISI 1018 mild steel was friction stir welded using a tungsten-based tool with the input parameters of tool rotational speed as 1000 rpm and welding speed as 50 mm/min. The paper presented fine-grained structures in the weld region that led to an increase in the tensile strength of the joint in comparison to the base metal. However, the toughness of the joints was decreased because of the induction of tungsten particles in the weld region (Lakshminarayanan et al., 2010).

A study to evaluate the correlation between the input parameters and microstructure evolution was conducted on a mild steel specimen. The outcome of the study was that the process parameters play a pivotal role in producing a defect-free joint. When the welding speed was kept high and rotational speed was kept low, then the heat input reduces; additionally, there was less flowability of the material, which resulted in defected welds, and the observable defect was the tunnel defect. On the other hand, if the welding speed was kept low and rotational speed was kept high, then there was an observable increase in heat input that led to improvement in the grain structure of the weldment and enhanced mechanical properties (Karami et al., 2016).

Figure 6.6 shows the optical macrographs of friction stir–welded high-strength low-alloy steel (HSLA) welded under different values of tool rotations using tungsten rhenium tool. The macrographs ensured that there was a formation of a successful joint at each level of rotational speed, and proper penetration was ensured in each situation. However, there was the formation of groove defects and tunnel defects in the weld region marked by arrows in Figure 6.6(e). The main reason for these defects was mainly due to lower heat input and inappropriate stirring action of the tool. The paper also presented different regions: weld nugget (WN), thermomechanical affected zone (TMAZ), and heat affected zone (HAZ), in the micrograph of the weld formed at 97 rpm, as shown in Figure 6.7. There micrograph shows that the TMAZ grains were oriented because of the shear stresses formed during the process (Ramesh et al., 2017).

Figure 6.8 represents the microstructural images of HSLA and friction stir welds. The major characterized portions are predominantly two-ferrites and pearlites. In

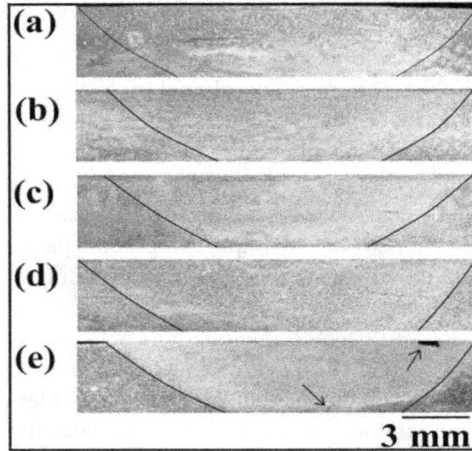

FIGURE 6.6 Optical macrograph of friction stir–welded joints at traverse speed of a) 57 mm/min, b) 67 mm/min, c) 77 mm/min, d) 87 mm/min, and e) 97 mm/min (Ramesh et al., 2017).

Source: © [Elsevier]. Reproduced by permission of Elsevier

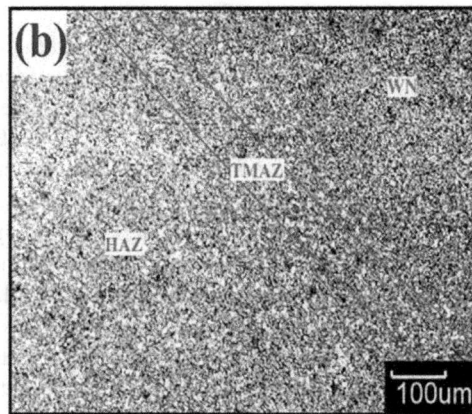

FIGURE 6.7 Optical micrograph of transition zone at a traverse speed of 97 mm/min (Ramesh et al., 2017).

Source: © [Elsevier]. Reproduced by permission of Elsevier

the micrograph, ferrite is represented by polygonal structures and colored bands, whereas pearlites are identified by the dark bands with fine structures. The major space in the micrograph is occupied by ferrites in comparison to pearlites. The reason for this is lower carbon content. The micrographs for the weld nugget are shown in Figure 6.8(b–f), and in the grain refinement is significantly improved as compared

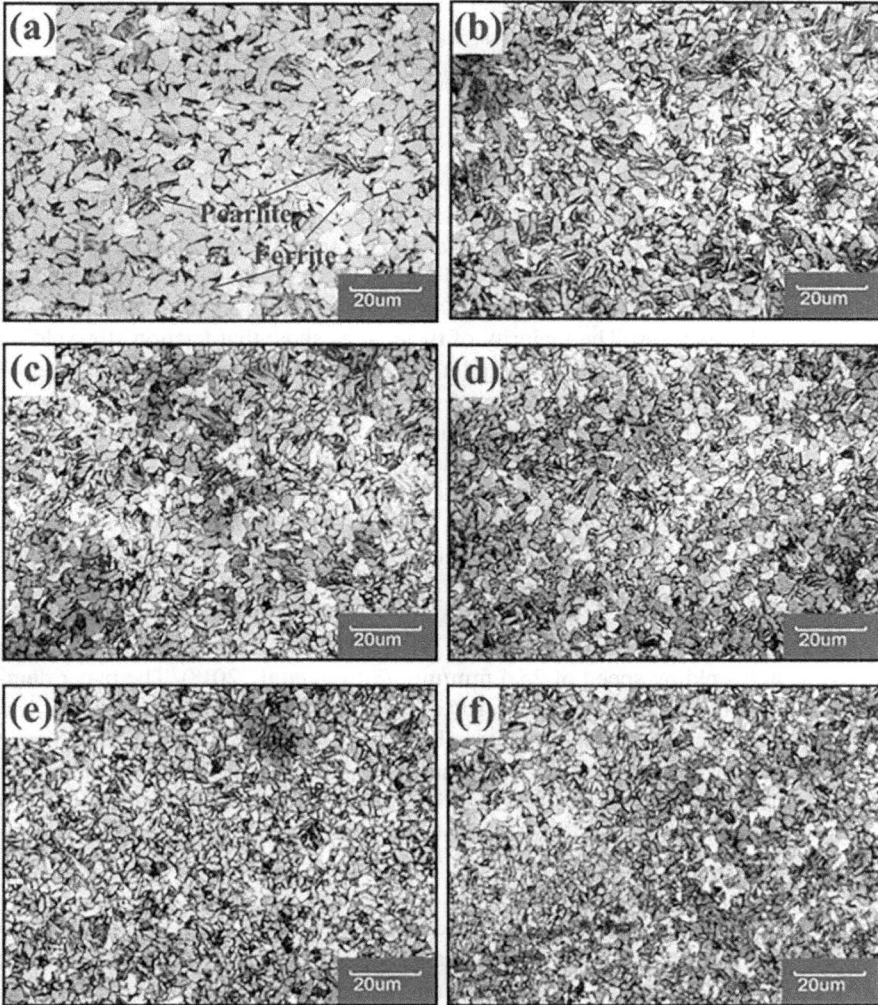

FIGURE 6.8 Optical macrograph of a) HSLA steel and friction stir–welded joints at a traverse speed of b) 57 mm/min, c) 67 mm/min, d) 77 mm/min, e) 87 mm/min, and f) 97 mm/min (Ramesh et al., 2017).

Source: © [Elsevier]. Reproduced by permission of Elsevier

to the base metal. The micrographs are not similar to each other because they are produced under different welding conditions.

Friction stir welding was conducted on X80 pipeline steel at different levels of tool rotational speeds: 300, 400, and 600 rpm. The lower tool speeds resulted in decreased levels of heat input, which in turn led to the formation of refined austenitic structures. As far as the weld nugget is concerned, there was the strongest shear texture at the level of medium heat input. The research outcome was that it was able to achieve an excellent weld toughness, which is nearly 93% of the base metal. The reason for this is low heat input and a higher packet boundaries ratio that consist of excellently refined martensitic and ferritic structures (Duan et al., 2020)

6.5 MECHANICAL PROPERTIES

The main objective of any joining process is to achieve a defect-free joint. The idea of the joint being defect free is attained by analyzing the mechanical properties of the welded joints such as tensile strength and hardness of the welds. There are numerous research papers available that stress the enhancement of mechanical properties of the welded structures. The majority of the studies show that friction stir welding is capable of producing a defect-free joint with improved mechanical properties as compared to the base metal.

Lean duplex stainless steel, 1.5 mm thick, was friction stir welded using a tungsten carbide tool at different welding speeds, 50, 100, and 150 mm/min (Esmailzadeh et al., 2013). The authors concluded that the change in welding conditions affects the mechanical properties of the welded joint. The paper reported that with an increase in welding speed, the grain size in the weld zone refines. Thus, there is an increase in the tensile strength and hardness values of the welded steel as compared to base metal.

APMT (Fe-22Cr-5Al-3Mo alloy) was friction stir welded with a tool rate of 600 RPM and welding speed of 25.4 mm/min (Sittiho et al., 2018). The paper demonstrated the feasibility of joining an iron alloy using the FSW process. Mechanical properties such as microhardness and tensile strength were tested. The results showed that there was no change in the microhardness values of the weld nugget as compared to the base metal. There was a significant improvement in the tensile strength of the stir zone in comparison with the base metal. The welding was carried out with a high value of welding speed that in turn led to an increment in strain rates and reduction in peak temperatures. In turn, the paper concluded that low heat input led to formation of fine grains, which improves the mechanical properties of the joint. Figure 6.9 shows a comparison between the yield and tensile strength of the parent metal and FSWed specimen for both the transverse and longitudinal directions.

Weathering mild steel was welded using friction stir welding at a low temperature value. The weathering steels considered for the experimentation were SMA490-AW and SPA-H joined at a temperature below A1 temperature (Wang et al., 2021). The joints produced by the FSW process exhibit two steps of yield because of high value of microhardness and strength of weldment. The joint strength exhibited by FSWed specimens was as high as 100%.

Friction stir welds of carbon steels (IF steel, S1C, and S35C) were prepared for microhardness characterization (Fujii et al., 2006). The hardness profiles are presented for a) tool RPM = 400 rpm and welding speed = 400 mm/min, b) tool RPM = 400 rpm and welding speed = 100 mm/min, and c) tool RPM = 400 rpm and welding speed = 200 mm/min, as shown in Figure 6.10. The base metal consisted of pearlite and cementite, and the weld region showed an increment in hardness. The stir zone showed increased hardness compared to the base metal and was almost

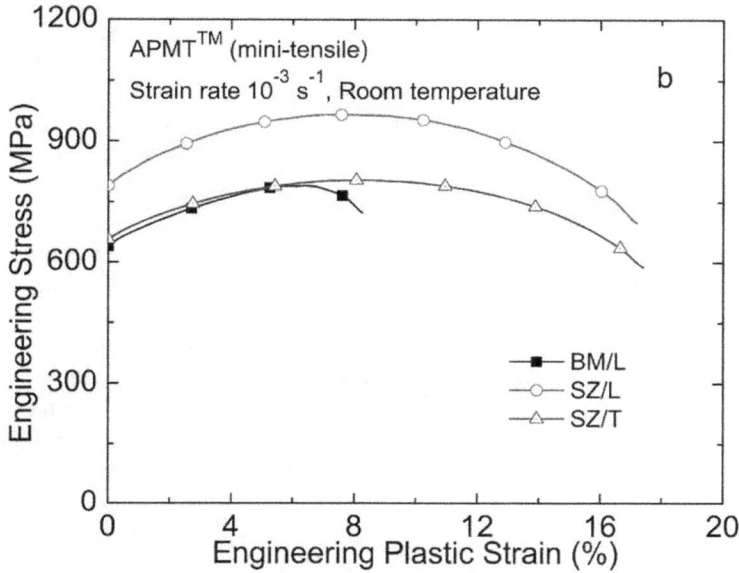

FIGURE 6.9 Engineering stress–engineering plastic strain curves of the as-received and FSWed APMT from a) full-length tensile specimen testing in the transverse direction (T); and b) mini-tensile specimen testing in both transverse (T) and longitudinal (L) directions (Sittiho et al., 2018).

Source: © [Elsevier]. Reproduced by permission of Elsevier

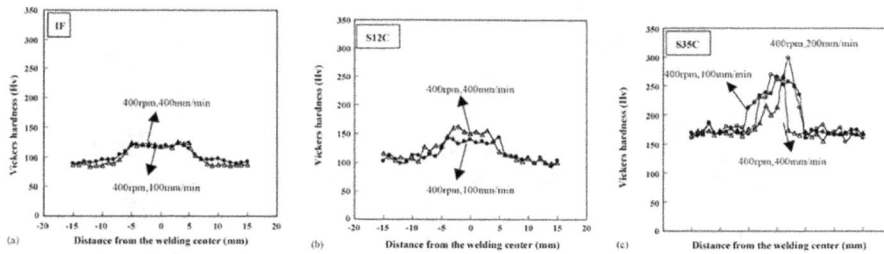

FIGURE 6.10 Microhardness profile of friction stir–welded carbon steels for: a) IF steel; b) S12C; and c) S35C (Fujii et al., 2006).

Source: © [Elsevier]. Reproduced by permission of Elsevier

homogeneous throughout the entire stir zone, as there was an increment in temperature (Esmailzadeh et al., 2013; Weinberger et al., 2009). This hardness increase is governed by the fact that there is very fine pearlite and martensite. The weld region had a high level of hardness because the weld nugget transformed into a region of refined grain structure (Legendre et al., 2009).

6.6 SURFACE DEFECTS

The most common defects that occur on the surface of friction stir–welded steels are surface cracks and grooves. The main reason for the occurrence of these types of defects is improper plasticization, which mainly occurs due to input parameters like tool RPM, plunge force, shoulder diameter, and welding speed. The different types of surface defects that occur during friction stir welding are illustrated in Figure 6.11. The majority of defects illustrated in Figure 6.11 are because of tool geometry or the flow of material (Arbegast, 2008). Defects that occur because of improper tool penetration are termed geometric defects, and defects that occur due to heat generation in the weld zone are termed flow defects. When the heat produced at the weld zone becomes very high, then the weld zone becomes plasticized beyond a certain limit, which results in the formation of a flash that leads to surface deformities and nugget collapse (Zhang et al., 2006). This inappropriate material flow can also occur when the speed of welding is very slow and may lead to the formation of fills on the surface, which leads to the formation of defects on the advancing side, forming a warm hole (Kumar & Kailas, 2008). Optimization of process parameters is critical in obtaining defect-free welds. The surface finish obtained in this process is governed by the process parameters like welding speed and tool RPM (Podržaj et al., 2015). As the tool rotates, it moves along the protruding surface, which leads to the formation of onion rings on the surface. The governing parameters also play a major role in achieving good mechanical properties of the welded joints. If welding is performed at a higher rate of welding speed or a lower level of tool rotation, then the stress concentrations can lead to deterioration in the mechanical properties. Surface groove is

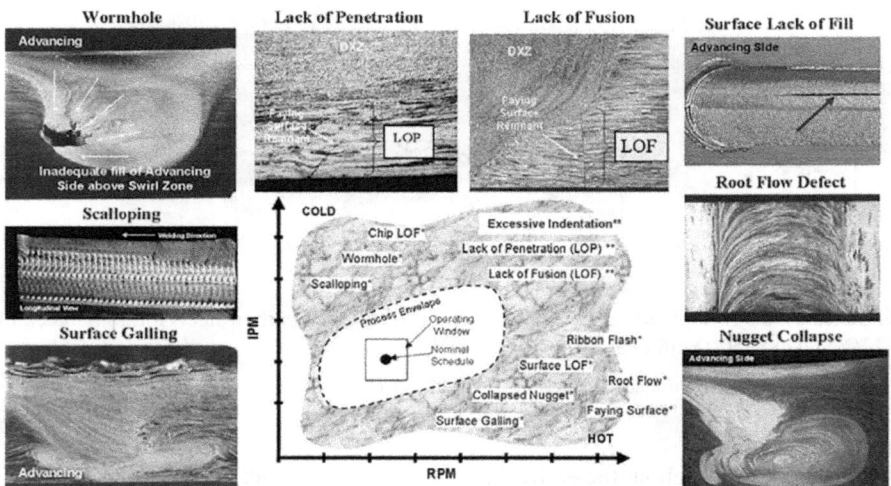

FIGURE 6.11 Types of defects produced during FSW of steels in context with material flow (Arbegast, 2008).

Source: © [Elsevier]. Reproduced by permission of Elsevier

one of the important defects that occur because of insufficient heat input and inappropriate plunge force (Kim et al., 2006)

6.7 APPLICATIONS OF FRICTION STIR WELDING

Friction stir welding is found to be applicable in the following areas.

6.7.1 SHIP BUILDING

Ships are subjected to extreme conditions such as cyclic loading with changing amplitudes because of dwell. These cyclic loadings may cause fatigue damage in naval structures, especially in welded joints (Fricke, 2003; Lautrou & Thevenet, 2005). These welded structures are often subjected to impact loadings and are also required to bear extreme conditions of temperature.

The fusion welding process, when applied to large steel plates with a thin gauge section, results in unacceptable levels of distortion in the final product that require expensive flame straightening or rework. In context with stainless steel, weld distortion is expected to be even greater than carbon steel because of lower thermal conductivity and higher thermal expansion. The *USS Arleigh Burke* (DDG-51)–class ship had an enormous cost for its construction in terms of flame straightening (Maria Posada et al., 2003). There are chances that defects such as porosity, cracks, and hydrogen embrittlement can occur in fusion welds. For naval applications, defects are very dangerous, as ships are subject to extreme conditions. Friction stir welding has come up with a solution, and the ship-building industry was the first to utilize friction stir welding.

6.7.2 ROAD TRANSPORT

Friction stir welding is also used in welding of automotive steels. One of the variants of FSW, friction stir spot welding, also is developing rapidly and can be very useful to weld automotive steels. Friction stir welding is being evaluated for its applications in various other industries.

6.7.3 NUCLEAR APPLICATIONS

The ability of friction-stir welding to produce defect-free welds and quality joints has gained the attention of nuclear industries (Cederqvist & Öberg, 2008). Friction stir welding is being assessed by several companies working in the nuclear sector, and they have reported improved mechanical properties (Müller et al., 2006; Pramann et al., 2014)

As far as nuclear applications are concerned, reduced-activation ferritic martensitic steels are widely used in nuclear reactors because of their properties. Friction stir welding of this steel was conducted to demonstrate the effects of change in mechanical properties of the welds. The welds were considered for testing for different temperature zones. The paper concluded that there was an increase in the yield strength of the material at different welding conditions. Therefore, this welded structure could be efficiently used in nuclear applications (Li et al., 2021)

Many applications of friction stir welding are coming up. The Welding Institute (TWI) the inventor of friction stir welding, is hopeful that this process will find applications in the following areas in the near future: electric motor housings, refrigeration panels, cooking equipment, gas tanks and cylinders, and so on.

6.8 CONCLUSION AND FUTURE SCOPE OF WORK

Based on the figures and facts presented in this work, it can be concluded that researchers have overcome challenges, and the welding of steels using friction stir welding has attained a new level of maturity. There were many experiments conducted in the recent past to find the optimum combination of tool type; tool profile; and process parameters like welding speed, tool RPM, shoulder diameter, and so on. This chapter highlights that a variety of steels can be welded using friction stir welding process with enhanced mechanical properties. In order to have an idea of structural changes, this process can achieve fine-grained structures in the weld region with a controlled cooling rate and welding temperature, thereby enhancing the mechanical properties like tensile strength, impact strength, and fatigue strength of the weld. The friction stir welding process has proved sustainable in comparison to the conventional fusion welding processes, as it offers enormous advantages: the process is a green welding process, as it does not create any environmental pollution; the process is cost effective, as no filler material is required; it does not lead to any hydrogen embrittlement; HAZ produced in this process is also smaller in context with fusion joining processes; and the joint strength and quality of the weld are comparatively higher as compared to fusion welded steel structures. There is a need to conduct more experimental research on joining steels using the FSW process to reduce the cost, which will facilitate the commercialization of FSW of steels. The interesting thing to conclude from the present chapter is that most of the research was conducted to enhance the joint strength of the welds, and less research was conducted on the economic aspects of joining.

REFERENCES

Arbegast, W. J. (2008). A flow-partitioned deformation zone model for defect formation during friction stir welding. *Scripta Materialia*, *58*(5), 372–376. https://doi.org/10.1016/j.scriptamat.2007.10.031

Ardalanniya, A., Nourouzi, S., & Jamshidi Aval, H. (2021). Fabrication of a laminated aluminium matrix composite using friction stir processing as a cladding method. *Materials Science and Engineering B: Solid-State Materials for Advanced Technology*, *272*(October 2020), 115326. https://doi.org/10.1016/j.mseb.2021.115326

Avettand-Fènoël, M. N., Nagaoka, T., Fujii, H., & Taillard, R. (2019). Effect of a Ni interlayer on microstructure and mechanical properties of WC-12Co cermet/SC45 steel friction stir welds. *Journal of Manufacturing Processes*, *40*(December 2018), 1–15. https://doi.org/10.1016/j.jmapro.2019.02.032

Avila, J. A., Rodriguez, J., Mei, P. R., & Ramirez, A. J. (2016). Microstructure and fracture toughness of multipass friction stir welded joints of API-5L-X80 steel plates. *Materials Science and Engineering A, 673*, 257–265. https://doi.org/10.1016/j.msea.2016.07.045

Barnes, S. J., Steuwer, A., Mahawish, S., Johnson, R., & Withers, P. J. (2008). Residual strains and microstructure development in single and sequential double sided friction stir welds in RQT-701 steel. *Materials Science and Engineering A, 492*(1–2), 35–44. https://doi.org/10.1016/j.msea.2008.02.049

Bhatia, A., & Wattal, R. (2021). Process parameters optimization for maximizing tensile strength in friction stir-welded carbon steel. *Strojniski Vestnik/Journal of Mechanical Engineering, 67*(6), 311–321. https://doi.org/10.5545/sv-jme.2021.7203

Cartigueyen, S., & Mahadevan, K. (2015). Role of friction stir processing on copper and copper based particle reinforced composites—A review. *Journal of Materials Science & Surface Engineering, 2*(2), 133–145.

Cederqvist, L., & Öberg, T. (2008). Reliability study of friction stir welded copper canisters containing Sweden's nuclear waste. *Reliability Engineering and System Safety, 93*(10), 1491–1499. https://doi.org/10.1016/j.ress.2007.09.010

Cho, H. H., Han, H. N., Hong, S. T., Park, J. H., Kwon, Y. J., Kim, S. H., & Steel, R. J. (2011). Microstructural analysis of friction stir welded ferritic stainless steel. *Materials Science and Engineering A, 528*(6), 2889–2894. https://doi.org/10.1016/j.msea.2010.12.061

Duan, R. H., Xie, G. M., Luo, Z. A., Xue, P., Wang, C., Misra, R. D. K., & Wang, G. D. (2020). Microstructure, crystallography, and toughness in nugget zone of friction stir welded high-strength pipeline steel. *Materials Science and Engineering A, 791*(March), 139620. https://doi.org/10.1016/j.msea.2020.139620

Elangovan, K., & Balasubramanian, V. (2008). Influences of tool pin profile and welding speed on the formation of friction stir processing zone in AA2219 aluminium alloy. *Journal of Materials Processing Technology, 200*(1–3), 163–175. https://doi.org/10.1016/j.jmatprotec.2007.09.019

Esmailzadeh, M., Shamanian, M., Kermanpur, A., & Saeid, T. (2013). Microstructure and mechanical properties of friction stir welded lean duplex stainless steel. *Materials Science and Engineering A, 561*, 486–491. https://doi.org/10.1016/j.msea.2012.10.068

Feng, A. H., & Ma, Z. Y. (2007). Enhanced mechanical properties of Mg-Al-Zn cast alloy via friction stir processing. *Scripta Materialia, 56*(5), 397–400. https://doi.org/10.1016/j.scriptamat.2006.10.035

Fricke, W. (2003). Fatigue analysis of welded joints: State of development. *Marine Structures, 16*(3), 185–200. https://doi.org/10.1016/S0951-8339(02)00075-8

Fujii, H., Cui, L., Tsuji, N., Maeda, M., Nakata, K., & Nogi, K. (2006). Friction stir welding of carbon steels. *Materials Science and Engineering A, 429*(1–2), 50–57. https://doi.org/10.1016/j.msea.2006.04.118

Gan, W., Li, Z. T., & Khurana, S. (2007). Tool materials selection for friction stir welding of L80 steel. *Science and Technology of Welding and Joining, 12*(7), 610–613. https://doi.org/10.1179/174329307X213792

Geiger, M., Micari, F., Merklein, M., Fratini, L., Contorno, D., Giera, A., & Staud, D. (2008). Friction stir knead welding of steel aluminium butt joints. *International Journal of Machine Tools and Manufacture, 48*(5), 515–521. https://doi.org/10.1016/j.ijmachtools.2007.08.002

Ghasemi-kahrizsangi, A., Kashani-Bozorg, S. F., & Moshref-Javadi, M. (2015). Effect of friction stir processing on the tribological performance of steel/Al_2O_3 nanocomposites. *Surface and Coatings Technology, 276*, 507–515. https://doi.org/10.1016/j.surfcoat.2015.06.023

Ghosh, M., Kumar, K., & Mishra, R. S. (2010). Analysis of microstructural evolution during friction stir welding of ultrahigh-strength steel. *Scripta Materialia*, *63*(8), 851–854. https://doi.org/10.1016/j.scriptamat.2010.06.032

Imam, M., Ueji, R., & Fujii, H. (2016). Effect of online rapid cooling on microstructure and mechanical properties of friction stir welded medium carbon steel. *Journal of Materials Processing Technology*, *230*, 62–71. https://doi.org/10.1016/j.jmatprotec.2015.11.015

Ji, Y. S., Fujii, H., Sun, Y., Maeda, M., Nakata, K., Kimura, H., Inoue, A., & Nogi, K. (2009). Friction stir welding of Zr55Cu50Ni 5Al10 bulk metallic glass. *Materials Transactions*, *50*(6), 1300–1303. https://doi.org/10.2320/matertrans.ME200806

Karami, S., Jafarian, H., Eivani, A. R., & Kheirandish, S. (2016). Engineering tensile properties by controlling welding parameters and microstructure in a mild steel processed by friction stir welding. *Materials Science and Engineering A*, *670*, 68–74. https://doi.org/10.1016/j.msea.2016.06.008

Karami, V., Dariani, B. M., & Hashemi, R. (2021). Investigation of forming limit curves and mechanical properties of 316 stainless steel/St37 steel tailor-welded blanks produced by tungsten inert gas and friction stir welding method. *CIRP Journal of Manufacturing Science and Technology*, *32*, 437–446. https://doi.org/10.1016/j.cirpj.2021.02.002

Ke, L., Huang, C., Xing, L., & Huang, K. (2010). Al-Ni intermetallic composites produced in situ by friction stir processing. *Journal of Alloys and Compounds*, *503*(2), 494–499. https://doi.org/10.1016/j.jallcom.2010.05.040

Kim, Y. G., Fujii, H., Tsumura, T., Komazaki, T., & Nakata, K. (2006). Three defect types in friction stir welding of aluminum die casting alloy. *Materials Science and Engineering A*, *415*(1–2), 250–254. https://doi.org/10.1016/j.msea.2005.09.072

Klingensmith, S., Dupont, J. N., & Marder, A. R. (2005). Microstructural characterization of a double-sided friction stir weld on a superaustenitic stainless steel. *Welding Journal*, 77–86.

Küçükömeroğlu, T., Aktarer, S. M., İpekoğlu, G., & Çam, G. (2018). Microstructure and mechanical properties of friction-stir welded St52 steel joints. *International Journal of Minerals, Metallurgy and Materials*, *25*(12), 1457–1464. https://doi.org/10.1007/s12613-018-1700-x

Kumar, K., & Kailas, S. V. (2008). The role of friction stir welding tool on material flow and weld formation. *Materials Science and Engineering A*, *485*(1–2), 367–374. https://doi.org/10.1016/j.msea.2007.08.013

Lakshminarayanan, A. K., Annamalai, V. E., & Elangovan, K. (2015). Identification of optimum friction stir spot welding process parameters controlling the properties of low carbon automotive steel joints. *Journal of Materials Research and Technology*, *4*(3), 262–272. https://doi.org/10.1016/j.jmrt.2015.01.001

Lakshminarayanan, A. K., Balasubramanian, V., & Salahuddin, M. (2010). Microstructure, tensile and impact toughness properties of friction stir welded mild steel. *Journal of Iron and Steel Research International*, *17*(10), 68–74. https://doi.org/10.1016/S1006-706X(10)60186-0

Lautrou, N., & Thevenet, D. (2005). A fatigue crack initiation approach for naval welded joints nicolas. *Oceans—Europe 2005*, *2*, 1163–1170. https://doi.org/10.1109/OCEANSE.2005.1513223

Legendre, F., Poissonnet, S., Bonnaillie, P., Boulanger, L., & Forest, L. (2009). Some microstructural characterisations in a friction stir welded oxide dispersion strengthened ferritic steel alloy. *Journal of Nuclear Materials*, *386–388*(C), 537–539. https://doi.org/10.1016/j.jnucmat.2008.12.170

Li, S., Yang, X., Vajragupta, N., Tang, W., Hartmaier, A., & Li, H. (2021). The influence of post-weld tempering temperatures on microstructure and strength in the stir zone of friction stir welded reduced activation ferritic/martensitic steel. *Materials Science and Engineering A*, *814*(March), 141224. https://doi.org/10.1016/j.msea.2021.141224

Lienert, T. J., Stellwag, W. L., Grimmett, B. B., & Warke, R. W. (2003). Friction stir welding studies on mild steel—process results, microstructures, and mechanical properties are reported. *The Welding Journal, Supplement*(January), 1–9.

Luo, J., Wang, X. J., & Wang, J. X. (2009). New technological methods and designs of stir head in resistance friction stir welding. *Science and Technology of Welding and Joining*, *14*(7), 650–654. https://doi.org/10.1179/136217109X12489665059429

Maria Posada, Jennifer P. Nguyen, David R. Forrest, J. J. D. (2003). Friction stir welding advances and joining technology. *AMP TIAC Quaterly (Special Issue)*, *7*(3), 18.

McNelley, T. R., Swaminathan, S., & Su, J. Q. (2008). Recrystallization mechanisms during friction stir welding/processing of aluminum alloys. *Scripta Materialia*, *58*(5), 349–354. https://doi.org/10.1016/j.scriptamat.2007.09.064

Miles, M. P., Nelson, T. W., Steel, R., Olsen, E., & Gallagher, M. (2009). Effect of friction stir welding conditions on properties and microstructures of high strength automotive steel. *Science and Technology of Welding and Joining*, *14*(3), 228–232. https://doi.org/10.1179/136217108x388633

Miles, M. P., Pew, J., Nelson, T. W., & Li, M. (2006). Comparison of formability of friction stir welded and laser welded dual phase 590 steel sheets. *Science and Technology of Welding and Joining*, *11*(4), 384–388. https://doi.org/10.1179/174329306X107737

Miles, M. P., Ridges, C. S., Hovanski, Y., Peterson, J., Santella, M. L., & Steel, R. (2011). Impact of tool wear on joint strength in friction stir spot welding of DP 980 steel. *Science and Technology of Welding and Joining*, *16*(7), 642–647. https://doi.org/10.1179/1362171811Y.0000000047

Mironov, S., Sato, Y. S., & Kokawa, H. (2008). Microstructural evolution during friction stir-processing of pure iron. *Acta Materialia*, *56*(11), 2602–2614. https://doi.org/10.1016/j.actamat.2008.01.040

Mishra, R. S., & Ma, Z. Y. (2005). Friction stir welding and processing. *Materials Science and Engineering R: Reports*, *50*(1–2), 1–78. https://doi.org/10.1016/j.mser.2005.07.001

Mishra, R. S., Ma, Z. Y., & Charit, I. (2003). Friction stir processing: A novel technique for fabrication of surface composite. *Materials Science and Engineering A*, *341*(1–2), 307–310. https://doi.org/10.1016/S0921-5093(02)00199-5

Muhammad, N. A., Wu, C., & Padhy, G. K. (2018). Review: Progress and trends in ultrasonic vibration assisted friction stir welding. *Journal of Harbin Institute of Technology (New Series)*, *25*(3), 16–42. https://doi.org/10.11916/j.issn.1005-9113.17105

Müller, C., Elaguine, M., Bellon, G., Ewert, U., Zscherpel, U., Scharmach, M., Reamer, B., Ryden, H., & Ronneteg, U. (2006). Reliability evaluation of NDT techniques for cu-welds for risk assessment of nuclear waste encapsulation. *Materialpruefung/Materials Testing*, *48*(3), 111–116. https://doi.org/10.3139/120.100199

Park, S. H. C., Sato, Y. S., Kokawa, H., Okamoto, K., Hirano, S., & Inagaki, M. (2003). Rapid formation of the sigma phase in 304 stainless steel during friction stir welding. *Scripta Materialia*, *49*(12), 1175–1180. https://doi.org/10.1016/j.scriptamat.2003.08.022

Podržaj, P., Jerman, B., & Klobčar, D. (2015). Welding defects at friction stir welding. *Metalurgija*, *54*(2), 387–389.

Pramann, Z., Thompson, B., Chrzanowski, J., & Mennel, D. (2014). Friction stir welding for a nuclear fusion reactor. *Materials Science Forum*, *783–786*, 1808–1813. https://doi.org/10.4028/www.scientific.net/msf.783-786.1808

Ramesh, R., Dinaharan, I., Kumar, R., & Akinlabi, E. T. (2017). Microstructure and mechanical characterization of friction stir welded high strength low alloy steels. *Materials Science and Engineering A, 687*, 39–46. https://doi.org/10.1016/j.msea.2017.01.050

Santella, M. L., Engstrom, T., Storjohann, D., & Pan, T. Y. (2005). Effects of friction stir processing on mechanical properties of the cast aluminum alloys A319 and A356. *Scripta Materialia, 53*(2), 201–206. https://doi.org/10.1016/j.scriptamat.2005.03.040

Sato, Y. S., Nelson, T. W., Sterling, C. J., Steel, R. J., & Pettersson, C.-O. (2005). Microstructure and mechanical properties of friction stir welded SAF 2507 super duplex stainless steel. *Materials Science and Engineering A, 397*, 376–384. https://doi.org/10.1016/j.msea.2012.10.068

Siddiquee, A. N., & Pandey, S. (2014). Experimental investigation on deformation and wear of WC tool during friction stir welding (FSW) of stainless steel. *International Journal of Advanced Manufacturing Technology, 73*(1–4), 479–486. https://doi.org/10.1007/s00170-014-5846-z

Sittiho, A., Tungala, V., Charit, I., & Mishra, R. S. (2018). Microstructure, mechanical properties and strengthening mechanisms of friction stir welded Kanthal APMT steel. *Journal of Nuclear Materials, 509*, 435–444. https://doi.org/10.1016/j.jnucmat.2018.07.001

Song, K. H., Tsumura, T., & Nakata, K. (2009). Development of microstructure and mechanical properties in laser-FSW hybrid welded inconel 600. *Materials Transactions, 50*(7), 1832–1837. https://doi.org/10.2320/matertrans.M2009058

Sued, M. K., Pons, D., Lavroff, J., & Wong, E. H. (2014). Design features for bobbin friction stir welding tools: Development of a conceptual model linking the underlying physics to the production process. *Materials and Design, 54*, 632–643. https://doi.org/10.1016/j.matdes.2013.08.057

Sun, Y. F., Konishi, Y., Kamai, M., & Fujii, H. (2013). Microstructure and mechanical properties of S45C steel prepared by laser-assisted friction stir welding. *Materials and Design, 47*, 842–849. https://doi.org/10.1016/j.matdes.2012.12.078

Sunil, B. R., Reddy, G. P. K., Patle, H., & Dumpala, R. (2016). Magnesium based surface metal matrix composites by friction stir processing. *Journal of Magnesium and Alloys, 4*(1), 52–61. https://doi.org/10.1016/j.jma.2016.02.001

Suryanarayanan, R., & Sridhar, V. G. (2020). Process parameter optimisation in pinless friction stir spot welding of dissimilar aluminium alloys using multi-start algorithm. *Proceedings of the Institution of Mechanical Engineers, Part C: Journal of Mechanical Engineering Science, 234*(20), 4101–4115. https://doi.org/10.1177/0954406220919482

Tiwari, A., Pankaj, P., Biswas, P., Kore, S. D., & Rao, A. G. (2019). Tool performance evaluation of friction stir welded shipbuilding grade DH36 steel butt joints. *International Journal of Advanced Manufacturing Technology, 103*(5–8), 1989–2005. https://doi.org/10.1007/s00170-019-03618-0

Ueji, R., Fujii, H., Cui, L., Nishioka, A., Kunishige, K., & Nogi, K. (2006). Friction stir welding of ultrafine grained plain low-carbon steel formed by the martensite process. *Materials Science and Engineering A, 423*(1–2), 324–330. https://doi.org/10.1016/j.msea.2006.02.038

Wang, D. A., Chao, C. W., Lin, P. C., & Uan, J. Y. (2010). Mechanical characterization of friction stir spot microwelds. *Journal of Materials Processing Technology, 210*(14), 1942–1948. https://doi.org/10.1016/j.jmatprotec.2010.07.005

Wang, W., Han, P., Peng, P., Zhang, T., Liu, Q., Yuan, S. N., Huang, L. Y., Yu, H. L., Qiao, K., & Wang, K. S. (2020). Friction stir processing of magnesium alloys: A review. *Acta Metallurgica Sinica (English Letters), 33*(1), 43–57. https://doi.org/10.1007/s40195-019-00971-7

Wang, W., Zhang, S., Qiao, K., Wang, K., Peng, P., Yuan, S., Chen, S., Zhang, T., Wang, Q., Liu, T., & Yang, Q. (2020). Microstructure and mechanical properties of friction stir welded joint of TRIP steel. *Journal of Manufacturing Processes, 56*(13), 623–634. https://doi.org/10.1016/j.jmapro.2020.05.045

Wang, Y., Tsutsumi, S., Kawakubo, T., & Fujii, H. (2021). Microstructure and mechanical properties of weathering mild steel joined by friction stir welding. *Materials Science and Engineering A, 823*(July), 141715. https://doi.org/10.1016/j.msea.2021.141715

Weinberger, T., Enzinger, N., & Cerjak, H. (2009). Microstructural and mechanical characterisation of friction stir welded 15–5PH steel. *Science and Technology of Welding and Joining, 14*(3), 210–215. https://doi.org/10.1179/136217109X406956

Yabuuchi, K., Tsuda, N., Kimura, A., Morisada, Y., Fujii, H., Serizawa, H., Nogami, S., Hasegawa, A., & Nagasaka, T. (2014). Effects of tool rotation speed on the mechanical properties and microstructure of friction stir welded ODS steel. *Materials Science and Engineering A, 595*, 291–296. https://doi.org/10.1016/j.msea.2013.12.022

Zhang, H., Lin, S. B., Wu, L., Feng, J. C., & Ma, S. L. (2006). Defects formation procedure and mathematic model for defect free friction stir welding of magnesium alloy. *Materials and Design, 27*(9), 805–809. https://doi.org/10.1016/j.matdes.2005.01.016

7 Recent Developments in Heat Treatment of Friction Stir–Welded Magnesium Alloy Joints

Kulwant Singh

CONTENTS

SUMMARY

> The greatest threat to our planet is the belief that someone else will save it.
>
> —Robert Swan

Protection of the environment is not only excellent for businesses, but it is also becoming an essential facet of how products are produced, procured, and administered, according to the world's most famous manufacturing organizations. Consequently, it is critical to think about sustainability at every stage of the manufacturing process. In addition to modernizing and improving current operations, it will be essential to planning new technologies that utilize minor hazardous elements and produce fewer emissions, qualifying them as green methods. Mg alloys for lightweight applications in the automotive sector are also a remarkable initiative to reduce emissions and conserve energy. This review highlights how the FSW method can assist in reducing the environmental impact of Mg alloy welding and points researchers to the right way to achieve the objective of sustainable practices. The attempts made by researchers to enhance the performance of FSW joints of Mg alloys using PWHT are also a significant addition to sustainable manufacturing.

DOI: 10.1201/9781003269298-7

7.1   INTRODUCTION

The strength-to-weight ratio makes Mg alloys imperative materials for weight reduction exercises. Mg alloys are a substitute for heavier material (cast iron, steels, copper alloys, etc.) day by day (Polmear et al., 2017; Singh et al., 2018a). Metal-matrix composites, aluminum, and magnesium can be used extensively in lightweight applications (Cole & Sherman, 1995). Most commercial Mg alloys are ternary, consisting of rare-earth elements, thorium, zinc, and aluminum. The main alloying component is aluminum in Mg-Al (ternary) categories that consist of AZ (Mg-Al-Zn), AM (Mg-Al-Mn), and AS (Mg-Al-Si) alloys (You et al., 2017). It is usual to classify magnesium alloys based on their use at ambient and higher temperatures. In high-temperature alloys, rare metals (earth) and thorium are used as alloying elements, whereas zinc and aluminum are employed as alloying components in room-temperature alloys (Avedesian & Baker, 1999). ASTM regulations are employed to recognize these Mg alloys. The first two letters in a series represent the critical code for key alloying constituents (as presented in Table 7.1). The concentration of these primary alloying elements is indicated by these two letters, preceded by two numerals. The fifth letter signifies the alloy modification. The temper designation of Mg alloys is comparable to the temper description of Al alloys (Polmear, 1994; Singh, 2019). Table 7.2 shows the fundamental temper designations for Mg alloys.

TABLE 7.1
The Principal Codes for Primary Alloying Elements

Constituent	Character
Aluminum	A
Antimony	Y
Bismuth	B
Copper	C
Chromium	R
Cadmium	D
Iron	F
Lithium	L
Lead	P
Manganese	M
Nickel	N
Rare earth	E
Silver	Q
Silicon	S
Thorium	H
Tin	T
Yttrium	W
Zirconium	K
Zinc	Z

Source: Avedesian & Baker, 1999; Polmear, 1994.

TABLE 7.2
Magnesium Alloys' Primary Temper Designations (continued)

Treatment	Designation	Use
		General Divisions
Thermal treatment to produce stable tempers other than F, O, or H	T	One or more numerals always precede T, and after heat treatment, products have stable tempers.
Solution heat-treatment, unstable temper	W	When the ageing time is specified, W is employed, and the alloy solution is heat-treated before ageing at room temperature to produce an unstable temper.
Strain-hardening	H	H is always preceded by two or more numerals, and wrought items are work hardened to achieve the appropriate ductility at reduced strength.
Annealing, recrystallization (wrought products only)	O	Wrought semis annealed to attain the least-strength temper and castings to improve ductile nature and dimension consistency.
As fabricated	F	No control over heat treatment or work hardening at elevated temperatures and hot working, cold working, and casting procedures used to fabricate.
		Subdivisions of the H Tempers
Strain-hardening and then stabilizing	H3, plus one or more digits	H3 is preceded by numerals that indicate the amount of work hardening that remains after stabilization and the attributes of work hardened items that have been stabilized. To keep these products from weakening with age, they must have their qualities stabilized.
Strain-hardening and then partially annealing	H2, plus one or more digits	H2 is preceded by numerals that represent residual work hardening after restricted annealing and partially annealed work-hardened products.
Strain-hardening only	H1, plus one or more digits	H1 is preceded by numerals indicating the degree of work hardening and work-hardened items that have not been exposed to excessive temperatures.
		Subdivisions of the T Tempers
Cooling, artificially ageing, and cold working	T10	The influence of cold working in flattening or straightening is recognized in property limits, and products are cold treated to improve strength.
Solution heat-treatment and ageing (artificially), followed by cold working	T9	After artificial aging, work hardening to enhance the strength.
Solution heat-treatment and cold work, followed by artificial ageing.	T8	Working at ambient temperature following solution treatment improves the product's strength. The flattening or straightening effects can be detected by product attributes.
Solution heat-treatment and stabilization	T7	Wrought products aged artificially. After the solution has been heated to its maximum strength, this procedure is carried out.
Solution heat-treatment and ageing (artificially)	T6	After solution heat treatment, products are not cold worked. Mechanical property restriction does not account for cold working effects.

(Continued)

TABLE 7.2

Magnesium Alloys' Primary Temper Designations (continued)

Treatment	Designation	Use
		Subdivisions of the T Tempers
Cooling and ageing artificial	T5	At room temperature, products are not worked on. The impacts of ambient temperature flattening and straightening are not recognized by product attributes.
Solution heat-treatment	T4	At room temperature, products are not worked on. The impacts of ambient temperature flattening and straightening are ignored by product characteristics.
Solution heat treatment and then cold working	T3	After solution treatment, the temperature is working. The straightening/flattening effects are recognized by product properties.
Annealing (cast products only)	T2	Working at room temperature improves the strength of the product. The straightening/flattening effects are recognized by product attributes.
Cooling and naturally ageing	T1	At room temperature, products are not worked on. Product characteristics ignore the impacts of ambient temperature straightening and flattening.

Source: Avedesian & Baker, 1999; Gupta & Sharon, 2011.

Magnesium alloys are widely used for numerous applications. However, there's still an absence of an efficient and effective welding process/technique for joining these alloys (Cao et al., 2009). Therefore, significant research is dedicated to the welding technologies/methods of these alloys like gas metal arc welding (Dong et al., 2012), electron beam welding (Chi et al., 2008), resistance spot welding (Xiao et al., 2011), and laser welding (Wang et al., 2007); however, some issues, like hot cracking, partial melting zone, porosity defects, and residual stress, are generated by these typical welding techniques/processes (Kah et al., 2015). These defects significantly deteriorate the welding joint properties and limit the widespread use of these lightweight alloys in the automobile and aerospace industries (Singh et al., 2020). Hence, a solid-state process is the most effective option/method to avoid such issues of fastening lightweight alloys. Friction stir welding is a versatile technique capable of joining metals, alloys, and composites (Kadigithala & Vanitha, 2020), and there is no need for any filler material while using FSW. Therefore, the associated metallurgic issues can be avoided, and this method can attain a sound weld (Singh et al., 2018b). Figure 7.1 demonstrates the FSW process.

Friction stir welding of Mg alloys offers widespread opportunities in essential industries such as land transportation, aircraft, railways, shipping and marine, constructions, and many others. The applications of FSW of magnesium alloy in various industries reported in the earlier published literature are shown in Figure 7.2. Although friction stir welding is a solid-state welding technique that may produce a low-distortion weld with superiority during fabrication of a weld joint, significant levels of residual stresses are induced within the joint (Singh et al., 2021; Staron et al., 2004). Commin et al. (2009) stated that the Mg alloy

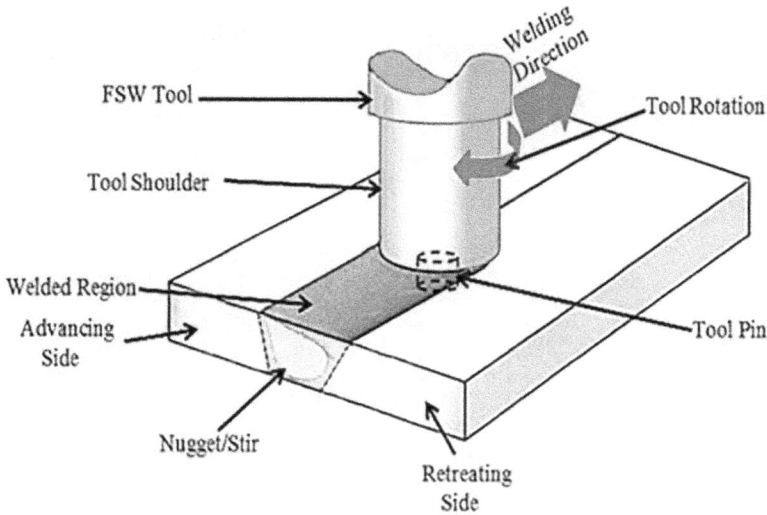

FIGURE 7.1 Friction stir welding illustration (Singh et al., 2019).

had restricted usage because of its high thermal conductivity, low melting point, active chemical nature, sensitivity to hot cracking, and complicated stresses (residual) within the weld zone. Commin et al. (2012) examined the impact on tensile characteristics of FSW joints of wrought Mg alloy. They observed the most considerable residue stresses (tensile type) in the joint's thermo-mechanically affected zone (TMAZ).

Ahmed et al. (2002) described that high residue stresses occurs in weld zones during FSW because of restraint by the base material on solidification. The prevailing literature shows that welding processes equally produced residual stresses because of the massive deformation in welding. Kouadri-Henni et al. (2016) described that the FSW method extremely changes the dispersal of residual stresses within the zones of the weldment. The heat-affected zone has weak residual (compressive) stresses, while the TMAZ and nugget zone (NZ) have residual (tensile) stresses. The profile of the residual stress in both of these zones (TMAZ and NZ) exhibited two peaks (noticeable) within the TMAZ. In addition, the profile transversal to the weld joint was found to be somewhat non-symmetrical. The thermo-mechanical affected zone did not show identical behavior on each side of the weld. Higher residual stresses have been observed in the weld joint's advancing side. Yan et al. (2017) stated that there are several strategies to recover welds' mechanical properties, including selecting suitable welding methods, selecting proper welding factors, and adjusting penetration fraction and appropriate heat treatment. Heat treatment after welding is a practical approach to remove residue stress and recover the properties of welds (Lobanov et al., 2016). This chapter reviews FSW joints to help researchers better understand the post-welding heat treatment (PWHT) of magnesium alloys.

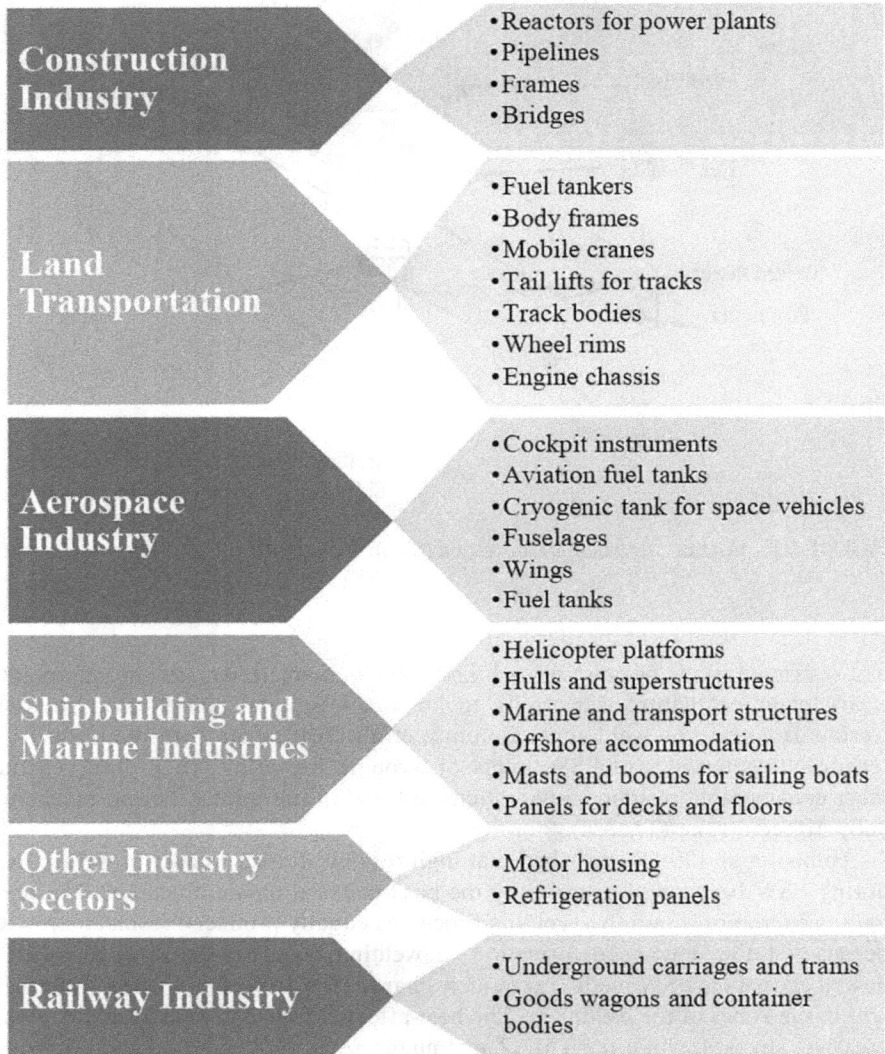

FIGURE 7.2 The applications of FSW of magnesium alloys (Lee et al., 2018; Prasad et al., 2021; Thomas & Nicholas, 1997).

7.2 POST-WELD HEAT TREATMENT

PWHT is a precise practice in which a weld is heated up to its lower critical transition temperature (specific) for a definite period (King, 2005). PWHT is useful for the removal of residual stresses and microstructure modifications that occurred during or after the welding process (Ott & Liu, 2018). To retain material properties after welding, PWHT is typically performed. PWHT is also used as a hardness control-ling technique to improve strength and reduce residual stresses (Funderburk, 1998). Ahmed & Krishnan (2002) reported that residual stresses occurred in weldments because of restraint within the base metal during solidification. These stresses could also be high because of the metal's yield strength level. When coupled with a normal force, these stresses may surpass the part's design stresses. The elemental concept of removal of residual stresses is that when the metal receives thermal energy, it per-mits the grain boundary to slip and eliminates metallurgical flaws (like vacancies, slip planes, and dislocations). The significant feature of PWHT is the avoidance of the material's brittle failure. Heat treatment improves the properties and softens the hardened zones to make it easy to machine. Removal of residual stresses introduced during or after welding is also essential for the dimensional stability of the weld joint (James, 1987). The choice of heat treatment method that involves stress-relief, nor-malizing, annealing, and aging depends upon the desired properties of the particular material (Totten, 2016). PWHT involves various potential treatments; however, post-heating and stress relieving are the broad types of heat treatment (Funderburk, 1998).

1. **Post-Heating**: During welding, extreme levels of hydrogen (ambient) spread throughout into a metal/material cause hydrogen-induced cracking (HIC) (King, 2009). The hydrogen is often dispersed from the welded space by heating the welded material to prevent hydrogen-induced cracking. This method is known as post-welding heating. The joint is heated to a definite temperature and retained at this temperature (according to the thickness and material nature) for a precise time, then allowed to cool down (Bryson, 2015).
2. **Stress Relieving**: During the welding method, residual stresses induced within the material causes corrosion (stress) and hydrogen-induced cracking (Beidokhti et al., 2009). These residual stresses are often reduced by post-welding heat treatment. This method constitutes heating the workpiece/metal to a specific temperature followed by step-by-step cooling (Jaluria, 2018; Lyons, 2014).

Some post-weld heat treatment studies of FSW Mg alloys are reported in the exist-ing literature (Afrinaldi et al., 2018; Fukumoto et al., 2007; Lin et al., 2015; Liu et al., 2018; Singh et al., 2019, 2020, 2021; G. Wang et al., 2017; L. Wang et al., 2017; Yang et al., 2014). Singh et al. (2020) examined the PWHT influences on FSW AZ91 Mg alloys. PWHT was performed at 260 °C for a soaking time of 1 hour. Singh et al. (2021) heat-treated FSW AZ31 Mg alloy joints for 15 min at 260 °C to deter-mine the importance of post-welding heat treatment. Yan et al. (2017) performed heat treatment of FSW AZ31B Mg alloy joints. The annealing was performed at 200 °C, 250 °C, 300 °C, 350 °C, 400 °C, and 450 °C for 60 minutes and solutionizing at 450 °C, 400 °C, 350 °C, 300 °C, and 250 °C for 10 hours, followed by cooling in the air at room temperature. Wang et al. (2017) explored the impact of PWHT on microstructural as well as mechanical properties of FSW AZ31 Mg alloy joints. The heat treatment was performed at 150 °C, 200 °C, 250 °C, 300 °C, 350 °C, 400 °C, and 450 °C for 60 minutes in an vacuum and followed by air cooling. Liu et al. (2018)

studied post-weld heat treatment of FSW joints of ZK60 Mg alloy. The samples were directly aged at 175 °C for 10 hours. Afrinaldi et al. (2018) accomplished post-aging treatment on friction stir processed plates of AMX602 Mg alloy at 180 °C for 40 hours. Lin et al. (2015) carried out post-welding heat treatment of AZ91 Mg alloys joints at 380 °C, 350 °C, and 320 °C for 60 minutes to 5 hours. Fukumoto et al. (2007) studied the heat treatment of FSW AZ31B joints at 500 °C, 400 °C, 300 °C, and 200 °C for a quarter-hour to 20 hours. Yang et al. (2014) conducted three heat treatments on the FSW joints of AZ80 magnesium alloy: (1) post-welding aging at 180 °C for 16 hours, (2) pre-welding solution treatment at 410 °C for 6 hours, and (3) pre-welding solution treatment and post-welding aging at 180 °C for 16 hours. Wang L. et al. (2017) studied post-weld heat treatment of Mg-Gd-Y-Zr alloy at 430 °C (solution treatment) for 4 hours followed by hot water quenching at 70 °C and ageing once more at 225 °C for 18 hours. Singh et al. (2019) performed PWHT on dissimilar Mg alloys joints at 260 °C for 15 min, half an hour, and 1 hour. The materials, temperature range, and soaking time used for post-weld heat treatment of Mg alloys within the existing literature are given in Table 7.3.

7.3 MICROSTRUCTURAL EVOLUTION

FSW modifies the microstructure of the parent material and leads to the development of stirring, thermo-mechanical affected, and heat-affected zones (Khaled, 2005), as shown in Figure 7.3.

Mishra & Ma (2005), Sahu & Pal (2018), and Cam G. (2011) also found that the FSW approach generates a variety of microstructural zones, each with its own set of microstructural properties. Heat treatment has a major impact on the microstructural zones and their characteristics (precipitate size and distribution, residual stress, dislocation density, and grain size) of the weld joint. Therefore, investigators must figure out the effects of PWHT of the joint. Singh et al. (2020) noticed that the grain structure in HAZ had changed and that new grains had emerged. After PWHT, the joint's ductile behavior improved. Singh et al. (2021) reported that the grain structure of the stirring zone had enhanced slightly, and grains had recrystallized. G. Wang et al. (2017) conveyed the consequences of heat treatment on the grain structure within the heat-affected zone. The grain size was more homogeneous when treated at 300 °C for 1 hour. Once the temperature went beyond 350 °C, new grains relocated to the grain boundary and accelerated recrystallization. At 450 °C, recrystallization was completed in a short time; therefore, the grain gave an impression of anomalous growth; a massive cluster grain appeared in some areas. Liu et al. (2018) reported that aging increases precipitates, which ends in a vital modification in the yield strength. Figure 7.4 shows the impact of post-welding treatments on the grain structure of ZK60 joints. J. Yang et al. (2014) described that the banded structure of AZ80 Mg alloy was observed within the stirring zone of the FSW joint because of different aluminum concentrations. After 16 hours of post-weld aging at 180 °C, the phase precipitated from the solution (supersaturated) in the parent metal in two ways: continuous and discontinuous precipitation (presented in Figure 7.5). Singh et al. (2019) conveyed that recrystallization of grains began after 15 and 30 minutes of heat treatment, but the grain structure could not fully recrystallize, resulting

TABLE 7.3

Summary of the Material, Temperature Range, and Soaking for Post-Weld Heat Treatment of Mg Alloys

Workpiece Material	Heat Treatment	Temperature Range (°C)	Soaking Time	Remarks	Reference
AZ91 Mg alloy	Artificial aging	260	60 Min	Improvement of the grain structure of weld zones.	(Singh et al., 2020)
AZ31 Mg Alloy	Artificial aging	260	15 Min	Strength (tensile) and elongation improved. Homogenized grains and enhanced mechanical properties.	(Singh et al., 2021)
AZ31 Mg alloy	Annealing and solutionizing	Annealing—200, 250, 300, 350, 400, 450 Solutionizing—250, 300, 350, 400, 450	Annealing—1 h Solutionizing—10 h	The homogeneous microstructure and the grains are much finer in the weld.	(Yan et al., 2017)
AZ31 Mg alloy	Aging in a vacuum and then air cooled	150, 200, 250, 300, 350, 400, 450	1 h	300 °C for 1 hour enhanced mechanical properties and homogenized grains.	(G. Wang et al., 2017)
ZK60 Mg alloy	Aging	175	10 h	The joint strength improved.	(Liu et al., 2018)
AMX602 Mg alloy	Aging	180	40 h	The hardness of the material increased.	(Afinaldi et al., 2018)
AZ91 Mg alloys	Aging	320, 350, 380	1–5 h	The shear strength was improved.	(Lin et al., 2015)
AZ31B Mg alloy	Aging	200, 300, 400, 500	15 min to 20 h	Almost negligible influence on mechanical properties.	(Fukumoto et al., 2007)
AZ80 Mg alloy	Aging, solutionizing, solution treatment plus aging	Aging—180 Solutionizing—410	Aging—16 h Solutionizing—6 h	About 30% yield strength improvement after aging.	(Yang et al., 2014)
Mg-Gd-Y-Zr alloy	Solution treatment plus aging	Solution treated—430 (hot water quenching at 70 °C) Aging—225	Solutionizing—4 h Aging—18 h	Enhanced the high-temperature tensile properties.	(L. Wang et al., 2017)
AZ61 and AZ91 Mg alloys	Artificial aging	260	15 min, 30 min, 60 min	Grain structure and tensile properties improved for 60 min.	(Singh et al., 2019)

FIGURE 7.3   Optical microstructure view of joint (Singh et al., 2016).

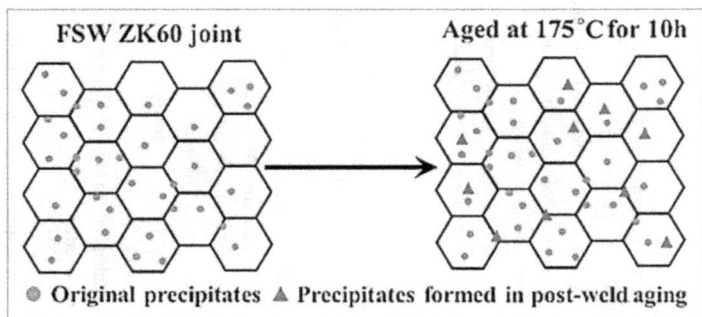

FIGURE 7.4   Precipitation during PWHT (Liu et al., 2018).

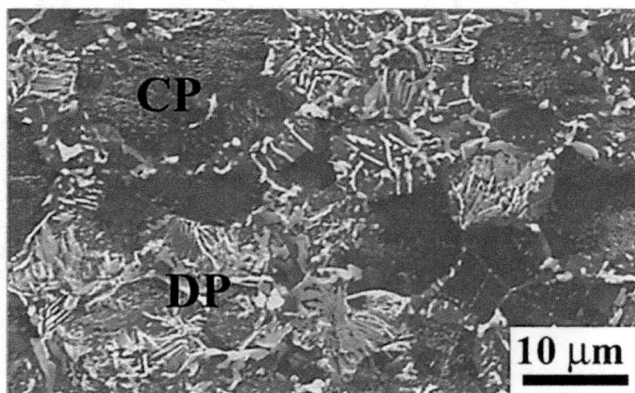

FIGURE 7.5   Continuous and discontinuous precipitation on post-weld aging (Yang et al., 2014).

in irregular grain development. However, the grain structure of the weldment area improved after 60 minutes of treatment. Here, the critical issue in microstructure evolution is acknowledging the refinement of grains and precipitation during the particular PWHT technique. The variation in the microstructural characteristics of the joint throughout PWHT depends on the specific PWHT technique, temperature range, and soaking time.

7.4   PROPERTIES

7.4.1   Residual Stresses

According to existing literature, massive deformation produces residual stresses throughout the welding processes. Just in the case of FSW, tensile (residual) stresses were observed in the longitudinal direction in the TMAZ. However, residual compressive stresses occurred transversely within the FSW zone (Kouadri-Henni et al., 2016; Lim et al., 2018; Silva et al., 2017). As FSW is a solid-state metal welding technique that may produce low-distorted weld joint of quality, considerable stresses (residual) are accomplished within the weld after fabrication (Commin et al., 2012; Staron et al., 2004). Ahmed & Krishnan (2002) conjointly described that residual stresses occurred within the weld joint of the material. Kouadri-Henni et al. (2016) conveyed that the FSW method amends stress distribution (residual) in the welding zones. The heat-affected zone has weak stresses (residual) of compressive nature; however, TMAZ and SZ have tensile nature residual stresses.

Because the profile of these stresses is slightly asymmetrical transverse to the weld, the TMAZ's behavior on the advancing and retreating sides of the weld has not been identified. The advancing side of the weld had a higher level of residual stress than the retreating side. Yan et al. (2017) conveyed several welding methods that may improve joints' mechanical properties. One can choose the suitable welding methodology, welding process parameters, penetration fraction, and heat treatment to get the desired properties of the joint. As reported earlier, post-welding heat treatment is an efficient technique to enhance the performance of the joints by removing the induced stress and recovering the property. G. Wang et al. (2017) reported that heat treatment could be a quite ancient method that is employed to eliminate residual stresses. These stresses are eliminated or adjusted by the thermal residual stress relaxation impact. By heat treatment at 300 °C for 60 minutes, they ascertained the significant changes in the tensile longitudinal residual stress as well as the compressive transversal residual stress of joints (presented in Figure 7.6). The critical literature reports that the stresses induced in the course of welding have to be taken away for better performance of the joint during the service, and these stresses are often easily removed by PWHT.

7.4.2   Mechanical Properties

According to the known literature, the PWHT causes considerable microstructural alterations in the FSW zones (Muruganandam, 2018; Thirumalvalavan et al., 2014). This variation in microstructure also disturbs the mechanical characteristics of the joint.

7.4.3   Tensile Properties

Heat treatment after welding has been shown to improve joint characteristics and soften the joint, making it easier to machine. Singh et al. (2020) reported 12.6 %

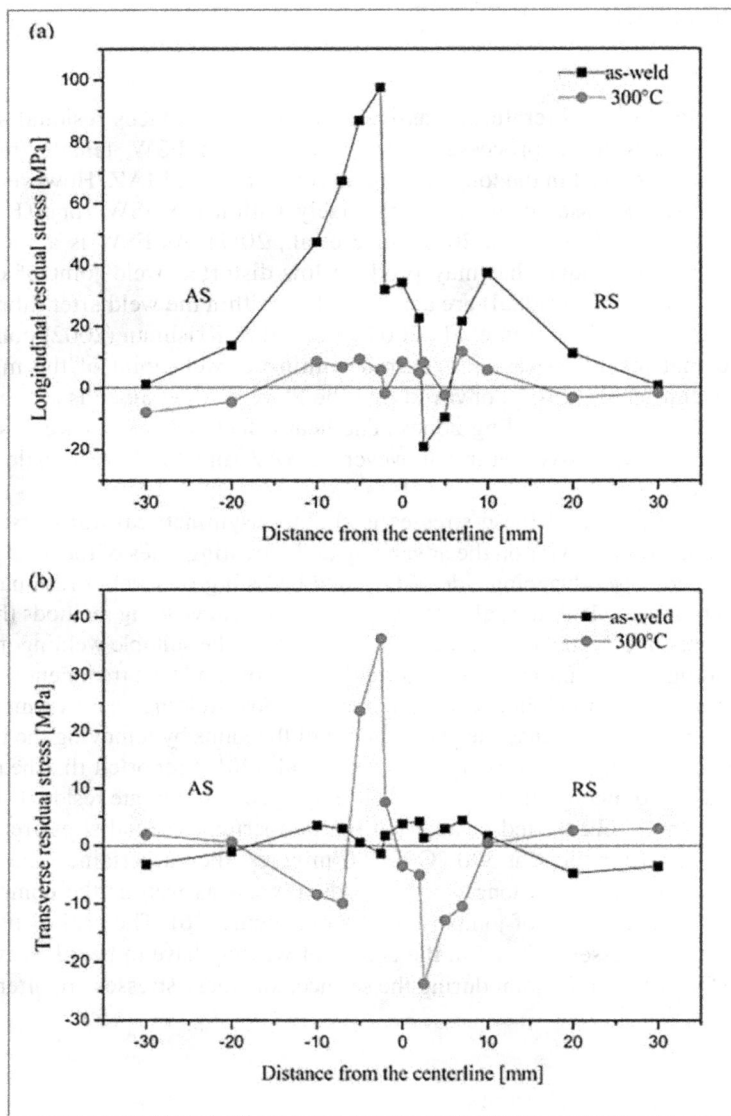

FIGURE 7.6 Residual stress distributions: a) longitudinal stress; b) transverse stress (G. Wang et al., 2017).

and 31.9 % improvement in the joint's tensile strength and elongation, respectively, after heat treatment. Singh et al. (2021) conveyed that the joint's strength (tensile) and percentage elongation improved by 4.7 % and 15.7 %, respectively, after PWHT. G. Wang et al. (2017) reported that the PWHT improved the tensile properties of welding joints; however, no substantial modification was observed in the base material's tensile properties, that is, the AZ31 Mg alloy of the heat treatment. The strength of the friction stir welded the PWHT enhanced joint at 300 °C for 60 minutes. The as-welded joint's yield strength (92.5 MPa) and tensile strength (199.1 MPa) were improved to 139.9 and 238.4 MPa, respectively. However, the heat treatment at 450 °C for 60 minutes further weakened the welded joint's strength (see Figure 7.7). Liu et al. (2018) reported the higher ultimate tensile strength (UTS) of the FSW joint (about 356 MPa) than FSW joints aged at 175 °C for 10 h (about 341 MPa). Fukumoto et al. (2007) reported that the tensile strength wasn't affected by PWHT, apart from a joint treated with 500 °C for 60 minutes. During the heat treatment at 500 °C, additional strength was removed (which produced work hardening) and resulted in lower tensile strength than the base metal AZ31B Mg alloy. Yang et al. (2014) reported that the post-weld aging of the AZ80 Mg alloy joint caused the enhancement of the yield strength (210 MPa, an increase of about 30%), a slight decrease in ultimate tensile strength (296 MPa), and a recognizable reduction in the elongation (about 5.7%). The grain refinement during PWHT may be a primary mode of improvement in the yield strength of the alloys. They reported that the precipitation of the β phase breaks the basal slip while aging. Therefore, cross slip and dislocation tangle induced increases in the yield strength dramatically and reduced ductility.

Heat treatment also altered FSW joints' fracture behavior. After post-FSW ageing, the fracture position migrated to the boundary of the stir zone, indicating that the

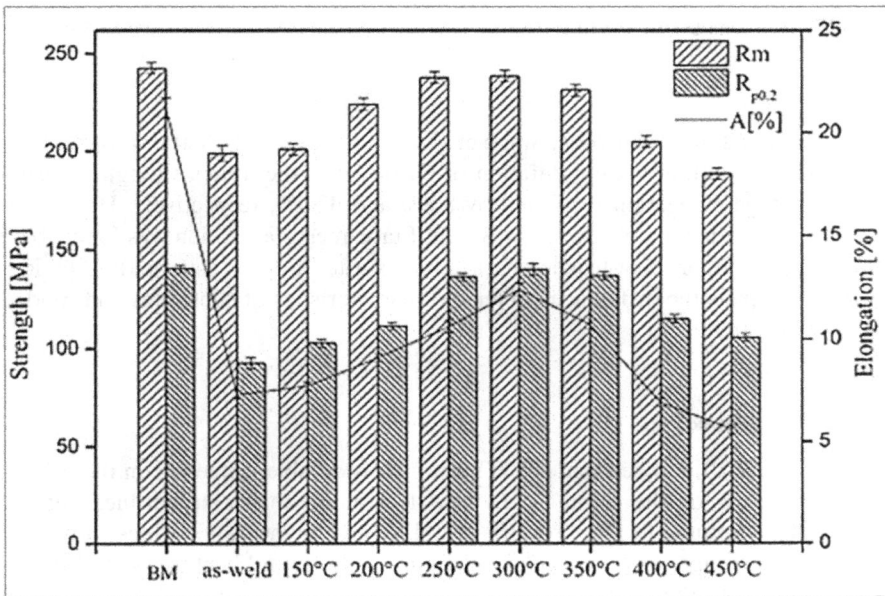

FIGURE 7.7 Tensile strength and elongation under heat treatment conditions (G. Wang et al., 2017).

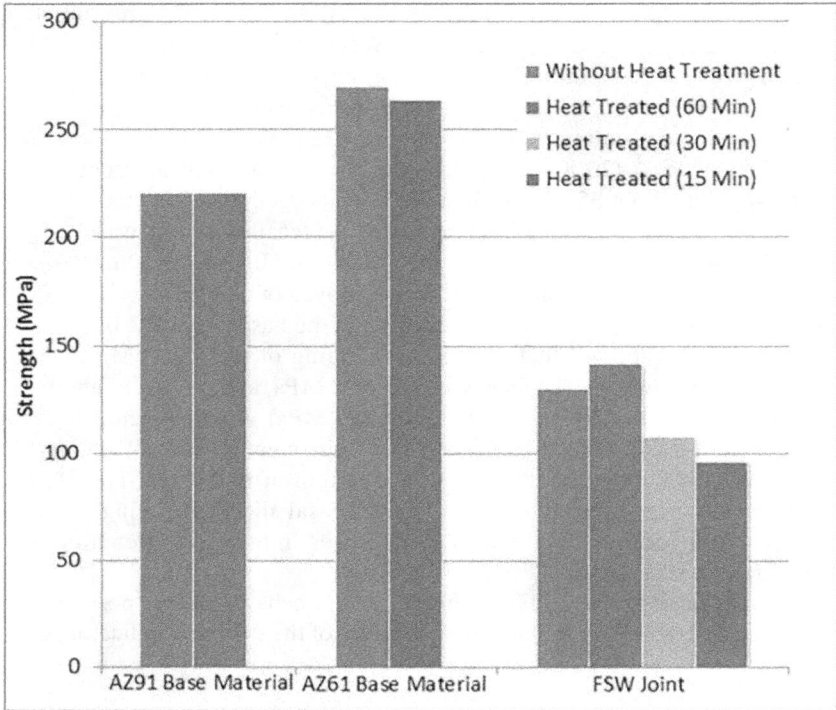

FIGURE 7.8 Tensile strength of substrates and welded joints on heat treatment (Singh et al., 2019).

FSW joint failed in the stir zone. Singh et al. (2019) conveyed that after post-welding heat treatment with a holding duration of 60 minutes, the joint's strength (tensile) and percentage elongation increased by 8.8% and 32.4%, respectively. Heat treatment increased the joint efficiency by 8.98%. Static recrystallization was found to be responsible for improving tensile characteristics after a 60-minute holding period. Figure 7.8 depicts the differences in tensile characteristics of substrates and welded joints on heat treatment.

7.4.4 MICROHARDNESS

Singh et al. (2020) noticed that PWHT has a considerable influence on the micro-hardness profile of the joint. After heat treatment, the microhardness of SZ decreased by around 11.11 % (from 78 to 70 Hv), and the overall hardness profile of the weld was comparably smooth. Singh et al. (2021) found heat treatment benefi-cial in smoothing the microhardness profile (reducing variations). Heat treatment enhanced the lowest microhardness in the heat-affected zone by about 2.5%. G.Wang et al. (2017) reported that the PWHT severely affected the microhardness distribution of the weld joint. They showed that every heat treatment technique had numerous effects on the microhardness profile of the FSW joint of AZ31 Mg alloy. In the weldment, the microhardness values observed were lower than that of parent metal (see Figure 7.9).

PWHT at 300 °C for 60 minutes refined and uniformly distributed the grains. They concluded that the refinement and uniform distribution of the grains during PWHT improved the hardness of the joint. J. Yang et al. (2014) described that after post-welding heat treatment (aging) of an AZ80 Mg alloy joint, the hardness of the parent metal and the stir zone were enhanced with the precipitation of the β phase. Because of the lower aluminum concentration in the matrix of the parent metal, hardness enhancement after post-weld aging was observed less in parent material than within the stirring zone. Singh et al. (2019) reported 12.95% lower hardness of the stir zone and an improvement of 4% in hardness in the advancing side after PWHT for 60 min, which improves the overall hardness profile of the FSW joints. Figure 7.10 illustrates the microhardness profile of FSW joints of dissimilar Mg alloys for different heat treatment conditions.

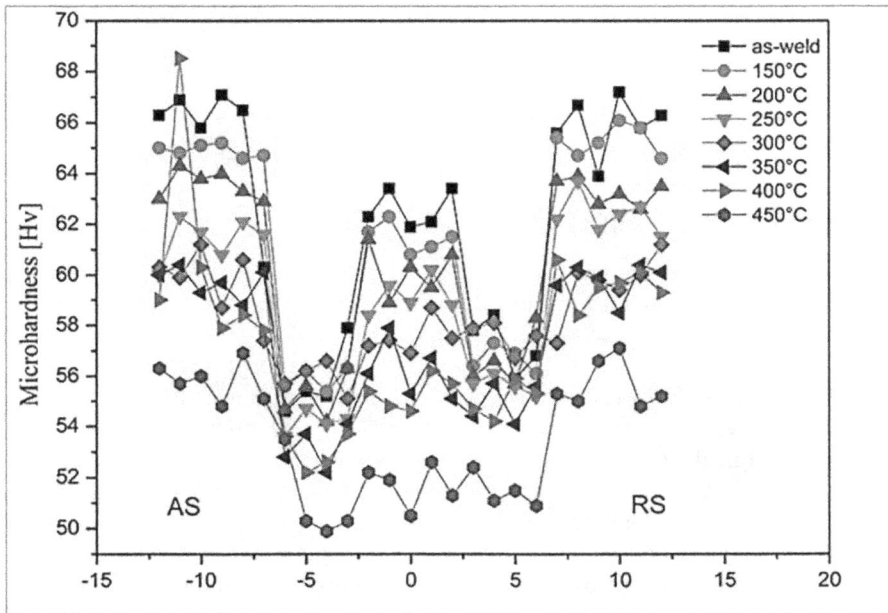

FIGURE 7.9   Microhardness profile of the joints (G. Wang et al., 2017).

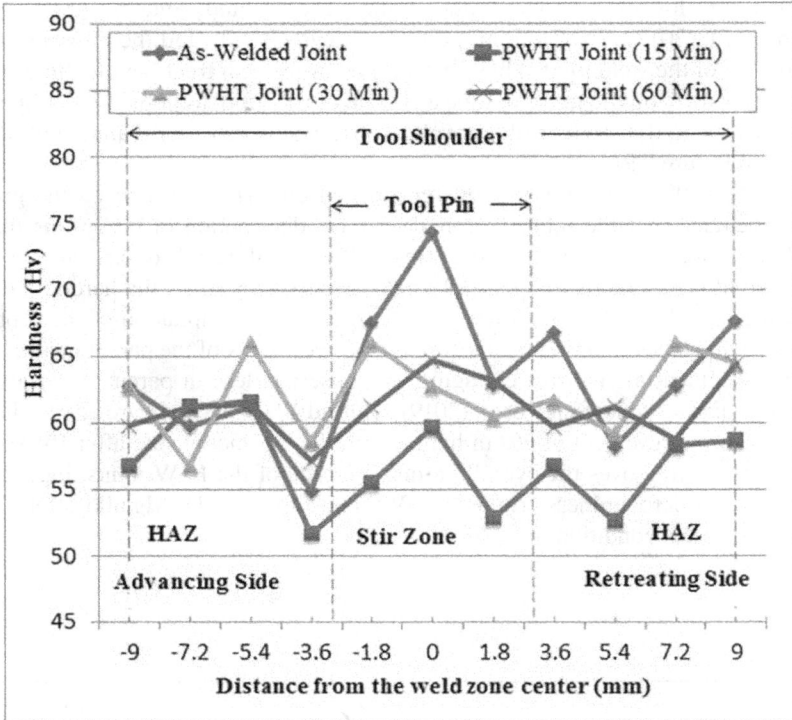

FIGURE 7.10 Microhardness profile of the joints (Singh et al., 2019).

7.5 CONCLUDING REMARKS

In this chapter, the current state of PWHT of magnesium alloys, including the effect of PWHT on grain structure, residual stresses, and mechanical characteristics of weld joints, has been reviewed. The conclusions are:

1. High stresses (residual) occurred in the weldment during solidification. These stresses could be as high as the metal's yield strength, resulting in metallurgical defects such as slip planes, vacancy, and dislocations, among other things. In this way, the residual stress induced in the weldment and weld exhibits brittle behavior.
2. Heat treatment is a helpful method for removing residual stress and restoring the characteristics of welded joints, hence improving joint performance. The removal of residual stresses is also essential for dimensional stability.
3. Post-welding heat treatment of the joint has a significant impact on the microstructural zones and their features.
4. Post-heating and stress relieving are the main two types of post welding heat treatment with unique procedures and objectives.

5. Natural aging (NA), solution treatment (ST), step aging (SA), solution treatment followed by artificial aging (STA), and artificial aging (AA) are the generally preferred techniques of PWHT.
6. The most significant feature of PWHT is the prevention of brittle fracture of the weld joint by giving a suitable treatment.
7. The aging period plays an imperative role during PWHT in improving joint efficiency.

7.6 FUTURE OUTLOOK

The available literature has well reported the heat treatment phenomena of weld joints with fruitful outcomes, but there is much scope for future research in this domain.

1. The various techniques (natural aging, solution treatment, step aging, solution treatment followed by artificial aging) and artificial aging of PWHT are employed by individuals. There is insufficient information in the literature, but selecting a particular technique for stress relieving or post-heating needs to be justified and generalized.
2. Generally, the effects of PWHT are reported in concerns of mechanical properties, residual stresses, and microstructure. Still, the texture of weldment is also affected during PWHT, so researchers should consider variation in texture during the heat treatment process.
3. The influence of welding settings and PWHT technique on thermal constancy of the joint's microstructure has to be better understood.
4. The elimination of stresses (produced during welding in the weldment) is a complicated phenomenon that needs to be fully understood.
5. Due to heating the material, corrosion behavior of material is also affected and still needs to be investigated.

REFERENCES

Afrinaldi, A., Kakiuchi, T., Itoh, R., Mizutani, Y., & Uematsu, Y. (2018). The effect of friction stir processing and post-aging treatment on fatigue behavior of Ca-added flame-resistant magnesium alloy. *International Journal of Advanced Manufacturing Technology*, 95(5–8), 2379–2391. https://doi.org/10.1007/s00170-017-1411-x
Ahmed, K., & Krishnan, K. (2002). Post-weld heat treatment—Case studies. *BARC Newsletter-Founder's Day Special Issue*, 111–115.
Avedesian, M. M., & Baker, H. (1999). ASM specialty handbook: magnesium and magnesium alloys. *ASM International*, 350.
Beidokhti, B., Dolati, A., & Koukabi, A. H. (2009). Effects of alloying elements and microstructure on the susceptibility of the welded HSLA steel to hydrogen-induced cracking and sulfide stress cracking. *Materials Science and Engineering: A*, 507(1–2), 167–173. https://doi.org/10.1016/j.msea.2008.11.064
Bryson, W. E. (2015). Heat Treatment Master Control Manual. In *Hanser*. Carl Hanser Verlag GmbH & Co. KG. https://doi.org/10.3139/9781569904862
Cam, G. (2011). Friction stir welded structural materials: beyond Al-alloys. *International Materials Reviews*, 56(1), 1–48. https://doi.org/10.1179/095066010x12777205875750

Cao, X., Jahazi, M., X. Cao, & M. Jahazi. (2009). Effect of welding speed on the quality of friction stir welded butt joints of a magnesium alloy. *Materials & Design*, *30*(6), 2033–2042. https://doi.org/10.1016/j.matdes.2008.08.040

Chi, C. T., Chao, C. G., Liu, T. F., & Wang, C. C. (2008). Relational analysis between parameters and defects for electron beam welding of AZ-series magnesium alloys. *Vacuum*, *82*(11), 1177–1182. https://doi.org/10.1016/j.vacuum.2007.12.019

Cole, G. S., & Sherman, A. M. (1995). Light weight materials for automotive applications. *Materials Characterization*, *35*(1), 3–9. https://doi.org/10.1016/1044-5803(95)00063-1

Commin, L., Dumont, M., Masse, J. E., & Barrallier, L. (2009). Friction stir welding of AZ31 magnesium alloy rolled sheets: Influence of processing parameters. *Acta Materialia*, *57*(2), 326–334. https://doi.org/10.1016/j.actamat.2008.09.011

Commin, Lorelei, Dumont, M., Rotinat, R., Pierron, F., Masse, J. E., & Barrallier, L. (2012). Influence of the microstructural changes and induced residual stresses on tensile properties of wrought magnesium alloy friction stir welds. *Materials Science and Engineering A*, *551*, 288–292. https://doi.org/10.1016/j.msea.2012.05.021

Dong, H. G., Liao, C. Q., & Yang, L. Q. (2012). Microstructure and mechanical properties of AZ31B magnesium alloy gas metal arc weld. *Transactions of Nonferrous Metals Society of China (English Edition)*, *22*(6), 1336–1341. https://doi.org/10.1016/S1003-6326(11)61323-X

Fukumoto, S., Yamamoto, D., Tomita, T., Okita, K., Tsubakino, H., & Yamamoto, A. (2007). Effect of post weld heat treatment on microstructures and mechanical properties of AZ31B friction welded joint. *Materials Transactions*, *48*(1), 44–52. https://doi.org/10.2320/matertrans.48.44

Gupta, M., & Sharon, N. M. L. (2011). Magnesium, magnesium alloys, and magnesium composites. In *Magnesium, Magnesium Alloys, and Magnesium Composites*. John Wiley & Sons, Inc. https://doi.org/10.1002/9780470905098

Jaluria, Y. (2018). *Advanced Materials Processing and Manufacturing*. Springer International Publishing. https://doi.org/10.1007/978-3-319-76983-7

James, M. R. (1987). Relaxation of residual stresses an overview. In *Residual Stresses* (pp. 349–365). Elsevier. https://doi.org/10.1016/B978-0-08-034062-3.50026-4

Kadigithala, N. K., & Vanitha, C. (2020). Investigation on the microstructure and mechanical properties of AZ91D magnesium alloy plates joined by friction stir welding. In *Lecture Notes in Mechanical Engineering* (pp. 1021–1030). Springer. https://doi.org/10.1007/978-981-15-1201-8_109

Kah, P., Rajan, R., Martikainen, J., & Suoranta, R. (2015). Investigation of weld defects in friction-stir welding and fusion welding of aluminium alloys. *International Journal of Mechanical and Materials Engineering*, *10*(1), 26. https://doi.org/10.1186/s40712-015-0053-8

Khaled, T. (2005). An outsider looks at friction stir welding. *Federal Aviation Administration*, 71.

King, B. (2005). *Welding and Post Weld Heat Treatment of 2.25%Cr-1%Mo Steel*. University of Wollongong. http://ro.uow.edu.au/theses/479

King, F. (2009). Hydrogen effects on carbon steel used fuel containers. *Nuclear Waste Management Organization*, 82.

Kouadri-Henni, A., Barrallier, L., & Badji, R. (2016). Residual stresses of a magnesium alloy (AZ31) welded by the friction stir welding processes. *MATEC Web of Conferences*, *80*, 06003. https://doi.org/10.1051/matecconf/20168006003

Lee, W. G., Kim, J.-S., Sun, S.-J., & Lim, J.-Y. (2018). The next generation material for lightweight railway car body structures: magnesium alloys. *Proceedings of the Institution of Mechanical Engineers, Part F: Journal of Rail and Rapid Transit*, *232*(1), 25–42. https://doi.org/10.1177/0954409716646140

Lim, Y.-S., Kim, S.-H., & Lee, K.-J. (2018). Effect of residual stress on the mechanical properties of FSW joints with SUS409L. *Advances in Materials Science and Engineering, 2018*, 1–8. https://doi.org/10.1155/2018/9890234

Lin, F., Tian, Y., Chen, Z., Wang, F., & Meng, Q. (2015). Diffusion bonding and post-weld heat treatment of extruded AZ91 magnesium alloys. *Medziagotyra, 21*(4), 532–535. https://doi.org/10.5755/j01.ms.21.4.9699

Liu, Z., Xin, R., Wu, X., Liu, D., & Liu, Q. (2018). Improvement in the strength of friction-stir-welded ZK60 alloys via post-weld compression and aging treatment. *Materials Science and Engineering A, 712*, 493–501. https://doi.org/10.1016/j.msea.2017.12.008

Lobanov, L. M., Pashchin, N. A., Mikhodui, O. L., & Khokhlova, J. A. (2016). Investigation of residual stresses in welded joints of heat-resistant magnesium alloy ML10 after electrodynamic treatment. *Journal of Magnesium and Alloys, 4*(2), 77–82. https://doi.org/10.1016/j.jma.2016.04.005

Lyons, A. (2014). Materials for architects and builders. In *Materials for Architects and Builders* (5th ed.). Routledge. https://doi.org/10.4324/9781315768748

Mishra, R. S., & Ma, Z. Y. (2005). Friction stir welding and processing. *Materials Science and Engineering: R: Reports, 50*(1–2), 1–78. https://doi.org/10.1016/j.mser.2005.07.001

Muruganandam, D. (2018). Influence of post weld heat treatment in friction stir welding of AA6061 and AZ61 Alloy. *Russian Journal of Nondestructive Testing, 54*(4), 294–301. https://doi.org/10.1134/S1061830918040095

Ott, E., & Liu, X. (2018). *Proceedings of the 9th International Symposium on Superalloy 718 & Derivatives: Energy, Aerospace, and Industrial Applications* (p. 1118). Springer International Publishing. https://doi.org/10.1007/978-3-319-89480-5

Polmear, I. J. (1994). Magnesium alloys and applications. *Materials Science and Technology, 10*(1), 1–16. https://doi.org/10.1179/mst.1994.10.1.1

Polmear, I., StJohn, D., Nie, J., & Qian, M. (2017). *Light Alloys: Metallurgy of the Light Metals*. Elsevier (Butterworth-Heinemann).

Prasad, S. V. S., Prasad, S. B., Verma, K., Mishra, R. K., Kumar, V., & Singh, S. (2021). The role and significance of Magnesium in modern day research-A review. *Journal of Magnesium and Alloys, xxxx*. https://doi.org/10.1016/j.jma.2021.05.012

Sahu, P. K., & Pal, S. (2018). Effect of FSW parameters on microstructure and mechanical properties of AM20 welds. *Materials and Manufacturing Processes, 33*(3), 288–298. https://doi.org/10.1080/10426914.2017.1279295

Scott Funderburk, R. (1998). Key concepts in welding engineering—Post weld heat treatment. *Welding Innovation, 15*(2). www.jflf.org/v/vspfiles/assets/pdf/keyconcepts4.pdf

Silva, E. P. da, Oliveira, V. B., Pereira, V. F., Maluf, O., Buzolin, R. H., & Pinto, H. C. (2017). Microstructure and residual stresses in a friction stir welded butt joint of as-cast ZK60 alloy containing rare earths. *Materials Research, 20*(3), 775–779. https://doi.org/10.1590/1980-5373-mr-2016-0899

Singh, K. (2019). *Experimental Studies on Friction Stir Welded Joints of Magnesium Alloy*. I. K. Gujral Punjab Technical University, Kapurthala.

Singh, K., Singh, G., & Singh, H. (2016). Friction stir welding of magnesium alloys: A review. *Asian Review of Mechanical Engineering, 5*(1), 5–8. www.trp.org.in

Singh, K., Singh, G., & Singh, H. (2018a). Investigation of microstructure and mechanical properties of friction stir welded AZ61 magnesium alloy joint. *Journal of Magnesium and Alloys, 6*(3), 292–298. https://doi.org/10.1016/j.jma.2018.05.004

Singh, K., Singh, G., & Singh, H. (2018b). Review on friction stir welding of magnesium alloys. *Journal of Magnesium and Alloys, 6*(4), 399–416. https://doi.org/10.1016/j.jma.2018.06.001

Singh, K., Singh, G., & Singh, H. (2019). Microstructure and mechanical behaviour of friction-stir-welded magnesium alloys: as-welded and post weld heat treated. *Materials Today Communications*, *20*, 100600. https://doi.org/10.1016/j.mtcomm.2019.100600

Singh, K., Singh, G., & Singh, H. (2020). Microstructural and mechanical behaviour evaluation of Mg-Al-Zn alloy friction stir welded joint. *International Journal of Automotive and Mechanical Engineering*, *17*(3), 8150–8159. https://doi.org/10.15282/ijame.17.3.2020.08.0612

Singh, K., Singh, G., & Singh, H. (2021). Influence of post welding heat treatment on the microstructure and mechanical properties of friction stir welding joint of AZ31 Mg alloy. *Proceedings of the Institution of Mechanical Engineers, Part E: Journal of Process Mechanical Engineering*, *235*(5), 1375–1382. https://doi.org/10.1177/0954408921997626

Staron, P., Kocak, M., Williams, S., & Wescott, A. (2004). Residual stress in friction stir-welded A1 sheets. *Physica B: Condensed Matter*, *350*(1–3 SUPPL. 1), E491–E493. https://doi.org/10.1016/j.physb.2004.03.128

Thirumalvalavan, S., Senthil, R., & Gnanavelbabu, A. (2014). Effect of heat treatment on tensile strength and microstructure of AZ61A magnesium alloy. *Applied Mechanics and Materials*, *606*(September 2013), 55–59. https://doi.org/10.4028/www.scientific.net/AMM.606.55

Thomas, W., & Nicholas, E. (1997). Friction stir welding for the transportation industries. *Materials & Design*, *18*(4–6), 269–273. https://doi.org/10.1016/S0261-3069(97)00062-9

Totten, G. E. (2016). *ASM Handbook Volume 4E: Heat Treating of Nonferrous Alloys* (Vol. 4). ASM International.

Wang, A. H., Xu, H. G., Yang, P., Zhang, X. L., & Xie, C. S. (2007). Nd: YAG laser butt welding of a 12 vol.% SiC particulate-reinforced magnesium alloy composite. *Materials Letters*, *61*(19–20), 4023–4026. https://doi.org/10.1016/j.matlet.2007.01.010

Wang, G., Yan, Z., Zhang, H., Zhang, X., Liu, F., Wang, X., & Su, Y. (2017). Improved properties of friction stir-welded AZ31 magnesium alloy by post-weld heat treatment. *Materials Science and Technology*, *33*(7), 854–863. https://doi.org/10.1080/02670836.2016.1243356

Wang, L., Huang, J., Dong, J., Li, Z., & Wu, Y. (2017). High temperature tensile properties of laser-welded high-strength Mg-Gd-Y-Zr alloy in as-welded and heat-treated conditions. *Welding in the World*, *61*(2), 299–306. https://doi.org/10.1007/s40194-016-0404-y

Xiao, L., Liu, L., Chen, D. L. L., Esmaeili, S., & Zhou, Y. (2011). Resistance spot weld fatigue behavior and dislocation substructures in two different heats of AZ31 magnesium alloy. *Materials Science and Engineering A*, *529*(1), 81–87. https://doi.org/10.1016/j.msea.2011.08.064

Yan, Z., Zhang, H., Duan, J., Liu, F., & Wang, G. (2017). Effect of post-weld heat treatment on mechanical characteristics of AZ31 magnesium alloy welded joints. *Journal Wuhan University of Technology, Materials Science Edition*, *32*(5), 1205–1212. https://doi.org/10.1007/s11595-017-1732-5

Yang, J., Ni, D. R., Wang, D., Xiao, B. L., & Ma, Z. Y. (2014). Friction stir welding of as-extruded Mg-Al-Zn alloy with higher Al content. Part II: Influence of precipitates. *Materials Characterization*, *96*, 135–141. https://doi.org/10.1016/j.matchar.2014.08.001

You, S., Huang, Y., Kainer, K. U., & Hort, N. (2017). Recent research and developments on wrought magnesium alloys. *Journal of Magnesium and Alloys*, *5*(3), 239–253. https://doi.org/10.1016/j.jma.2017.09.001

8 Benefits and Challenges in Additive Manufacturing and Its Applications

Rajwinder Singh, Mohammod Toseef,
Jaswinder Kumar, Jashanpreet Singh

CONTENTS

8.1 INTRODUCTION: BACKGROUND OF DEVELOPMENT IN MANUFACTURING TECHNOLOGIES

The manufacturing landscape is always evolving. This shift is being fueled by the development of new manufacturing technologies, which allow for more cost-efficient as well as resource-efficient micro-scale production. 3D printing has emerged as a direct manufacturing technique, in connection with trends like servitization (Neely, 2008), personalization (Zhou et al., 2013), and presumption (Fox & Li, 2012), which are prompting organizations to reconsider where and how they undertake manufacturing operations. Advanced manufacturing technologies like additive manufacturing (AM) and others seem to indicate that value chains of the future will be much shorter, more local, and collaborative, with significant environmental benefits (Gebler et al., 2014). Conventionally, the performance of industrial production systems was assessed by considering factors like quality, cost, flexibility, and time. However, sustainability as well as energy and resource efficiency now account for better performance of production systems; see Figure 8.1 (Chen et al., 2015). Assembling items layer by layer, additive manufacturing mimics the way the body works. When compared to typical subtractive manufacturing methods, it is fundamentally less wasteful and may be used to separate the production of social and economic value from the ecological effect of corporate activities. Three stand out among the numerous possible sustainability advantages of this technology:

- AM may be used to improve resource efficiency by modifying manufacturing processes and products, resulting in improvements in both production and usage phases.
- Refurbishment and repair, as well as more ecological socio-economic configurations like higher person-product empathies and tighter producer-consumer connections, may help extend the life of a product (Kohtala & Hyysalo, 2015).
- Examples of redesigned value networks include shorter and simpler supply chains, more localized manufacturing, novel distribution tactics, and new alliances.

Although these are potential advantages, AM is not been properly investigated from a sustainability standpoint. As an enabler and an improved industrial

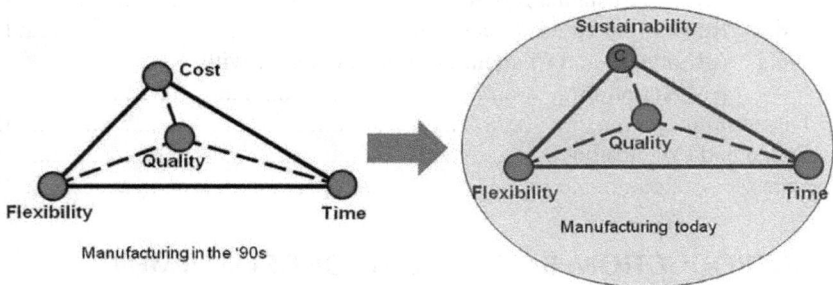

FIGURE 8.1 Performance assessment parameters of manufacturing/production systems in the 1990s and at present (Chen et al., 2015).

sustainability, it has the potential to lead to an alternative scenario where less eco-efficient localized manufacturing, consumer expectations for modified products, and a greater rate of product undesirability combine to raise resource consumption.

Since AM's inception, researchers have investigated the sustainability implications of the technology in a variety of ways, from broad to restricted (Baumers et al., 2011; Faludi et al., 2015; Gebler et al., 2014; Kohtala & Hyysalo, 2015). This chapter will summarize the following aspects:

- How might AM allow more sustainable methods of production and consumption?
- It is the purpose of this chapter to begin to unravel the challenges that exist at the crossroads of these themes by asking: How might additive manufacturing allow more sustainable forms of production and consumption?
- This is a new research field where the consequences of AM for sustainability remain unknown.

A deeper understanding of the consequences of AM for enhancing the sustainability of manufacturing systems may be gained by looking at the issue through the lens of industrial sustainability. "Goods (namely raw materials, different spare parts and products), services, and money" are all exchanged worldwide in complex and linked value chains. From digital production to peer-to-peer production, industrial systems include distributed manufacturing systems, including distributed production modes such as mass customization, customized fabrication, and mass production.

This chapter begins by offering an overview of additive manufacturing technology and its features, as well as a summary of its industry applications. This is followed by a comprehensive literature review of existing research on the subjects of AM's long-term viability. The chapter then goes over several instances in comparison to current practice. These examples demonstrate the various paths that can be taken, such as which companies have already started implementing AM, and what the results are and the impact of this technology's implementation on the long-term viability of the entire manufacturing process using the product life cycle as a guide. The categorization of these cases is based on one's perspective. Product and process redesign, as well as material input, are the four basic clusters: processing an component and product manufacture on a custom basis, as well as closing the loop. This chapter expands on these ideas by discussing how AM generates prospects for long-term sustainability and what kinds of opportunities AM creates.

8.2 REVIEW OF ADDITIVE MANUFACTURING TECHNOLOGY

One of several cutting-edge manufacturing techniques now under research, additive manufacturing has enormous potential to alter manufacturing distribution and society as a whole (Huang et al., 2013). Manufacturing technologies known as "additive

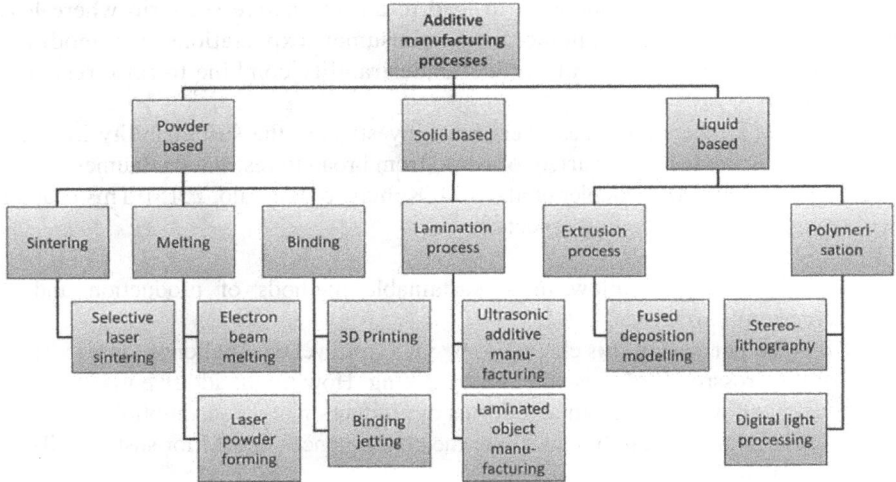

FIGURE 8.2   Various additive manufacturing processes (Chen et al., 2015).

manufacturing" may be used to create three-dimensional things that can be printed on demand. Various types of additive manufacturing processes are illustrated in Figure 8.2 (Chen et al., 2015). AM is a "method of mixing materials to manufacture products from 3D model data, often layer by layer, rather than subtractive manufacturing processes," according to the ASTM F42 Technical Committee (2012), which monitors the development of AM standards. Amplified materials are produced by a wide range of techniques, including fusion deposition modeling, stereolithography, and selective laser melting, as well as digital light-processing and selective laser sintering. Other AM techniques include: polyjets, electron beam melting, and laminated object manufacturing (Castro et al., 2015; Petrovic et al., 2011; Shamsaei et al., 2015). A variety of polymers, metals, ceramics, and composite materials may be used in AM. It all depends on the sort of additive manufacturing technique that is used to make these materials (Guo & Leu, 2013).

Rapid prototyping and tooling were the initial areas in which AM was used. However, as AM technologies continue to advance in terms of performance, they are rapidly used for direct manufacturing. Aerospace industries have long recognized the potential of additive manufacturing technologies and are investing in research to improve their reliability and application (Goh et al., 2017; Guo & Leu, 2013). Items that can only be used by a single person are essential in the medical industry. Because of its strengths, AM is the most effective strategy for satisfying this need. Hearing aids, such as in-the-ear models, are virtually entirely made in AM (Sandstrom, 2015), whereas other fields like as orthotics and dentures, implants, and replacement organs are in various phases of research and adoption. A typical AM-based 3D CAD model of core geometry of the dental bridge is shown in Figure 8.3 (Silva et al., 2017). It's easy to draw parallels between the current state of industrial development and earlier stages of industrial expansion and speciation (Ford et al., 2014; Phaal et al.,

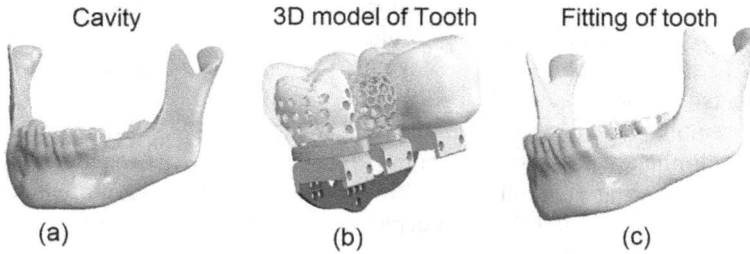

Cavity 3D model of Tooth Fitting of tooth

(a) (b) (c)

FIGURE 8.3 A typical AM-based 3D CAD model of core geometry of the dental bridge (Silva et al., 2017).

Source: Under the CC BY-NC-ND license http://creativecommons.org/licenses/by-nc-nd/4.0/

2011). Cold spray-based technology has made a lot of further breakthroughs since its inception.

8.3 ADDITIVE MANUFACTURING AND SUSTAINABILITY

Raw resources are transformed into final products and services during manufacturing. The environmental impact of production may be greatly influenced by the effectiveness of this conversion process (Gutowski et al., 2009). AM has been shown to provide a wide range of advantages involved in sustainability. To name a few advantages, the ability to optimize geometries and manufacture lightweight components reduces the amount of material used in production and consumption of energy during usage, as well as the amount of transportation used in the supply chain (Chen et al., 2015; Mani et al., 2014). This chapter predicts that AM can play an increasingly vital role in manufacturing in the future.

Few studies are currently looking into and assessing how well these potential benefits are being realized. In the most academic studies, AM has been compared to other manufacturing methods like injection molding (Baumers et al., 2011; Chen et al., 2015; Sreenivasan et al., 2010). These investigations have yielded a wide range of interesting results. A generalization about whether AM has a lower ecological impact as compared to other manufacturing techniques is difficult based on these studies because the impact on the life cycle of parts made with AM is decidedly reliant on machine utilization, the specification of AM equipment, and the input material processing method (Faludi et al., 2015). More study is needed in this area (Huang et al., 2013). However, the sophistication of machines is something that cannot be denied. Cutting costs requires the use of shared machines and tools.

AM is no exception to the rule that economics and the environment are intricately intertwined in manufacturing processes (Chen et al., 2015). It is possible to produce small to medium batches of metal components using AM technologies presently available. The overall cost is heavily influenced by the price per component of the machine. Although AM machines and materials are now pricey, they will become more affordable as the technology becomes more commonly used. AM is anticipated to become more economical in the future when larger manufacturing quantities are economically feasible. Product and component redesigns are possible since AM's design flexibility allows for this. With additive manufacturing, several components may be replaced by a single integrated assembly, decreasing or eliminating assembly-related costs, time, and quality difficulties. Consolidation of parts lowers or eliminates the cost of assembly. In order to achieve functional requirements while lowering material volume, it is possible to re-design a product. There is some evidence to suggest that AM deployment might reduce manufacturing costs significantly based on life cycle analyses. There will be a $113–370 billion savings by 2025 as a result of reduced input materials and processing. Materials used in additive manufacturing are not always more environmentally friendly than those used in conventional production, notwithstanding the possibility for better recycling rates in the latter. The various desired properties of materials used in AM are illustrated in Figure 8.4 (Daminabo et al., 2020; Liu et al., 2019). Polylactic acid (PLA),

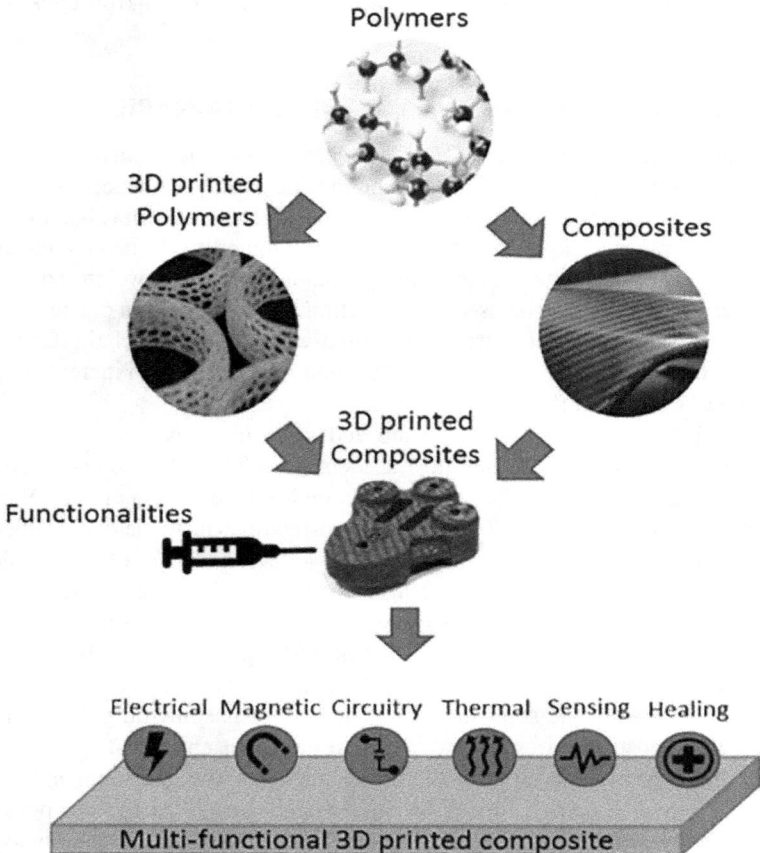

FIGURE 8.4 Various desired properties of materials used in AM (Daminabo et al., 2020; Liu et al., 2019).

a biopolymer, may be an exception to the rule (Faludi et al., 2015). It is possible that potential material savings are partly countered by the input material's relative toxicity (Faludi et al., 2015) and the impact of energy consumption during the process of generation of input material. According to some authors (Faludi et al., 2015; Hao et al., 2010; Sreenivasan et al., 2010), AM's overall environmental performance must take into consideration the system's energy consumption, not only the process's own shorter supply chains (Gebler et al., 2014).

Policymakers are starting to recognize AM's power to help create a more sustainable society. The UK-based Additive Manufacturing Special Interest Group (AM SIG) described how AM may satisfy future needs for sustainable, high-value production via a more effective manufacturing system and creative business models (TSB, 2012). AM will become an essential technology that allows design optimization, product lightness, inventory reduction, greater flexibility in the manufacturing site, products personalized to consumers, and an increase in the ability to produce goods that are unique to the individual consumer (BIS, 2013).

8.4   RESEARCH METHODS

However, researchers (Chen et al., 2015; Kohtala & Hyysalo, 2015; Mani et al., 2014) have outlined the possible advantages and dangers of additive manufacturing for sustainability, and it is crucial to look at the potential benefits and challenges that are applied in actual practice. This chapter examines the ramifications of AM adoption by businesses and sectors by conducting exploratory case studies (Yin, 2009). The examples were drawn from reports from consultants (e.g. Wohlers, Credit Suisse, PWC), recognized industry news sources (e.g. TCT Magazine, 3D Printing Industry, 3D Print Pulse, 3Ders), and industry experts. A typical example of fabrication of 3D printed composites from polymers is illustrated in Figure 8.5 (Bekas et al., 2019). Following the discovery of prospective examples, the selection of the cases for this study was guided by a life cycle viewpoint. To choose examples for comparing findings, theoretical replication logic was used. Because AM technologies are so new at the moment, the instances chosen are exemplars of how they may be used rather than typical of a wider population.

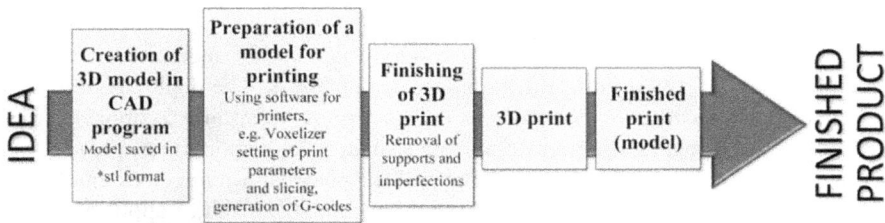

FIGURE 8.5   A typical example of fabrication of 3D printed composites from polymers (Bekas et al., 2019).

Source: Under the terms of the Creative Commons Attribution License 4.0 http://creativecommons.org/licenses/by/4.0/

8.5   AM AND SUSTAINABILITY IN INDUSTRY

This section explores the long-term consequences of AM's acceptance. The different phases of the product life cycle are discussed here.

8.5.1   PRODUCT AND PROCESS DESIGN

AM enables the construction of more sophisticated and optimized parts, as well as modest assemblies with fewer components and materials, because of its greater form and geometric freedoms.

8.5.2   COMPONENTS AND PRODUCT REDESIGN

Porous mesh arrays and open cellular foams may now be produced using the new AM design freedoms. Including these additional structures in a component's core may enhance the component's existing characteristics. The aerospace industry has found particular success with additive manufacturing because of the high performance demands and the restricted production scale (Al-Lami et al., 2018; Goh et al., 2017). The manufacturing of aircraft engine components has a substantial environmental effect. A considerable amount of waste is produced throughout the production process.

8.5.3   PRODUCT REDESIGN

In the same way that manufacturing process design may be improved, changes in product design can be made too. The use of AM technology-based components (e.g., molds, tooling) that use geometries exclusively possible via AM may help reduce the manufacturing process' energy and input material consumption (Chen et al., 2015). In order to effectively integrate AM systems into manufacturing processes and increase resource efficiency, they must be automated. Automated post-processing is needed to reduce the "stair stepping" impression caused by the slow layer-by-layer buildup of material.

8.5.4   BENEFITS AND LIMITATIONS

It is possible to boost both material and energy efficiency by reducing the amount of material and energy used in the manufacturing process. The deployment of AM may also lead to changes in the value chain. Redesigning items and components may lead to simpler things with fewer components, materials, personalities, phases, and interactions. However, in order to get the full advantages, a number of challenges and roadblocks must be overcome. Since AM technology's initial application was limited to rapid prototyping and tooling, designers and engineers have a distorted view of its suitability for direct component and product manufacture. Second, the performance of AM technologies is constrained. New functionality, such as microelectronics, cannot yet be incorporated into components or devices created using additive manufacturing.

8.5.5 MATERIAL INPUT PROCESSING

Sustainability may be improved by using AM inputs that are made from sustainable resources. Just as there is a wide variety of AM processes and materials, so there is a wide variety of inputs. The kind of additive manufacturing technology used determines the material's qualities. The four major materials are liquid, filament/paste, powder, and solid sheet (Guo & Leu, 2013). A rethinking of the raw material processing stage can help reduce the resources needed to bring some raw materials into useable form as inputs for manufacturing operations. The metal powders used in laser sintering and melting and electron beam melting are two such examples (Grossin et al., 2021). Metal ores need to be refined and processed before they can be used in production, which uses a lot of energy. Titanium powder can be made directly from titanium ore using a process developed by Metalysis, a UK-based start-up, which has commercialized the process (Lubik & Garnsey, 2016). For the purpose of making titanium powder, the FFC procedure requires significantly less energy than the more traditional Kroll procedure (Mellor et al., 2015). Calcium chloride is used as a nontoxic reactant during the refining process, and any leftover $CaCl_2$ can be recycled. A few materials have unique processing procedures, but this is due to the relatively young age of the technology. For those that have, there are no standards. An investigation into mechanical as well as thermal properties of materials is needed before this standardization can take place in AM technologies.

In the AM process, both recycled and new materials are used simultaneously. There are examples of home 3D printers that utilize fused deposition modeling (FDM) technology at the local level. A schematic procedure involved in the FDM process is shown in Figure 8.6 (Bryll et al., 2018). Errors, discarded plastic filament, and undesirable results may all be rescued and put to good use. Equipment like Filabots are used to attain this goal. The process starts with plastic objects being ground into granules, which are then fed into a filament-making machine. Because of the color contamination and degradation of polymer material properties, this procedure is not recommended. Polylactic acid, a polymer widely used in 3D printing filament and capable of being recycled without compromising quality, may help with the latter issue. At the system level, this is achieved by transforming regularly recycled materials into AM-ready shapes utilizing regularly recycled resources. Recycled polyethylene terephthalate (rPET) is used in the cartridges of the EKOCYCLE Cube home 3D printer. Red, white, black, and natural-colored cartridges are now available and include 25% recycled PET. Recycled materials can be used at higher concentrations, but the resulting polymer's visual appeal is limited. Using 3D systems and the Coca-Cola Company, a joint effort, this initiative is aimed at removing old PET bottles from landfills, such as those from Coca-Cola.

The last benefit of AM is that it may be used to turn waste and byproducts into new goods. There are several instances of luxury items being made out of recycled materials that were formerly regarded as rubbish. Because of the upcycling community (Braungart & McDonough, 2002), it is possible to generate something useful from something that would otherwise be deemed garbage (Braungart & McDonough, 2002). "Bewell Watches" provides some great instances of this work. Timber and woodworking produce a lot of powder and dust, which are discarded

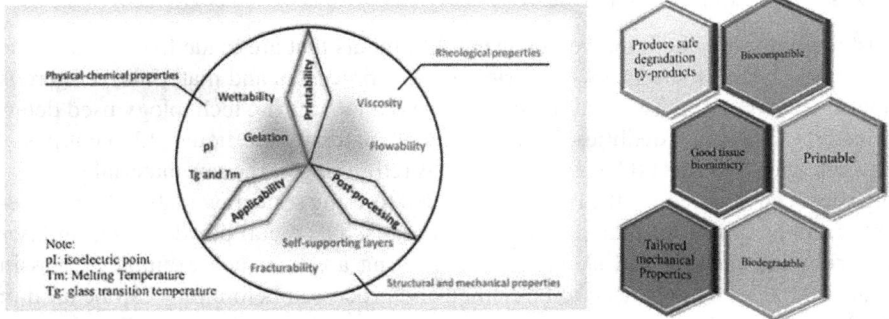

FIGURE 8.6 A schematic procedure involve in FDM process (Bryll et al., 2018).

as trash. Thermoset polymers, wood-plastic composites, and building goods all use these wood scraps as fillers. "Bewell Watches" creates wood fibers for additive manufacturing from waste wood by-products and adhesives, resulting in unique wooden clocks that provide value for both the firm and its clients. In the end, however, there are restrictions to material recirculation owing to quality and purity issues, which may prohibit items from being recycled. The recycling of wood-polymer composites is still in its infancy, and there are no national infrastructures in place to support it (Teuber et al., 2016).

8.5.6 Product Manufacturing

Economically, additive printing is the best method to manufacture unique components and products, allowing for the production of spare parts for replacement as well as lower-cost customization and personalization. In order to use additive manufacturing, you need a digital design library. One-off manufacturing of replacement parts may be more cost effective using AM, and 3D CAD files containing component designs can be readily shared after they've been manufactured. Access to certain types of material has been limited, however. A 3D CAD library provides access to replacement components that may be downloaded and built by the user. Kazzata also provides a service that links consumers in need of replacement parts with designers who can provide CAD design services, since the number of files available is presently restricted. Kazzata may provide the client with a 3D CAD file for self-manufacturing or the actual replacement part when it has been created. AM service bureaus may provide comparable design and engineering services as the number of AM service bureaus continues to grow. It's difficult to expand a wider system of spare parts because of concerns about component certification and legal responsibility.

For replacement components, Siemens Power Generation Services, a subsidiary of Siemens, is moving toward custom production (Siemens PGS). In its capacity as a subsidiary of Siemens AG, it offers technical support, maintenance, and repair services to customers that use rotating power equipment such gas and steam turbines, generators, and compressors. When Siemens PGS bought EOS DMLS AM machines in 2007, it cited the combustion system in gas turbines as an area where AM might improve customer value in spare parts maintenance and manufacturing. In order to take use of the design freedoms given by AM, Siemens PGS modified the burner "swirler" within this combustion system. AM can speed up the repair process and prevent excessive wastage. In comparison to the previous procedure, the repair time is expected to be ten times quicker. A lower amount of trash is generated since just the top 18 mm of the burner tip are removed prior to restoration. The company's long-term objective is to produce replacement parts on demand closer to the customer's location, and this is a step toward that goal by using AM.

In this part, we'll describe how the production system's configuration might change significantly when AM is introduced. Moving from conventional mass-production procedures and economies of scale to small batch production of customized or personalized products enables low-cost small-batch customization manufacturing. It is becoming more common for people and companies alike to employ 3D printers such as Ultimaker, Cube, and Makerbot Replicator in their own homes and enterprises. Prosumers are people who use technology to both generate and consume goods. These new technologies allow for the rapid development and production of goods in response to specific customer needs, as well as at the location where they will be used. The whole supply chain is projected to be significantly impacted by their broad deployment. Logistical processes are streamlined due to the utilization of fewer, but more essential, materials. Another benefit is reduced or eliminated inventories, which reduce economic losses and minimize environmental damage from obsolete components that are no longer in use (Chen et al., 2015).

AM's additive nature makes it a more environmentally friendly manufacturing approach than subtractive procedures since less waste is generated. Relative performance is a given, but make-to-order manufacturing enables AM to give superior absolute performance compared to traditional manufacturing methods, which spend more energy per unit produced. The greater usage of raw materials leads to dematerialization and waste reduction (Chen et al., 2015).

As a consequence of direct interaction between local customers and producers, the make-to-order method of distribution not only saves energy and resources but also encourages collaboration and user innovation (de Jong & de Bruijn, 2013). Examples include 3D hubs, which link 3D printer owners with potential customers through the internet. Prosumers are often the ones who own these printers, and they want to make greater use of the extra printing capacity they have. This creates new avenues for local manufacturing. It has the same advantages as those listed previously, but it also encourages the use of equipment by removing the need for customers to purchase and maintain their own. There has been a steady rise in the number of hubs in the network. At the time of this writing, there are around 25,000 3D printers in use in the 3D printing community.

Problems arise when actors with ambiguous roles and responsibilities collaborate in a non-linear, localized manner, leading to conflict (Chen et al., 2015; Petrick & Simpson, 2013). In addition, an ever-changing group of players and opponents creates an uncertain investment environment, making it difficult to define company strategies and competitive positions. For new businesses to succeed in this market, they need to be very robust, adaptable, and sensitive (Gartner, 1995).

8.5.7 CLOSING THE LOOP

Attempting to close the loop in AM may be done at various stages and sizes. The highest value recovery may be achieved when undesired AM material (powder or resin) is recovered throughout the manufacturing process. Metal powders can be recycled to a degree of 95–98%, according to current estimates (Petrovic et al., 2011). It is possible to recover more value from trash by using the AM approach. Polymers like PET, which are often used in consumer goods, may be repurposed into fashion accessories.

Public awareness and education on small-scale plastic recycling and additive manufacturing is a goal of initiatives like Better Future Factory. According to PPP, it is possible to use plastic waste as a 3D printing material. For everyday products like plastic cups and bottles as well as plastic bags at the grocery store, the materials tested include PLA, styrene (PS), low-density polyethylene (LDPE), polyester (PA), and polypropylene (PP). Even though certain polymers are more readily recycled than others, the experiment showed that recycling plastic for 3D printing applications is possible and very simple. Many plastics may be replaced with biopolymer PLA, for example, which has a broad variety of properties. The use of PLA and a smaller range of plastics might result in more straightforward recycling systems. When handled by experts, PLA may be recycled without sacrificing its quality (e.g., Plaxica). In a closed-loop system, it may be fed back into the original system (Chen et al., 2015).

Repair, maintenance, and remanufacturing may all benefit from the usage of the make-to-order strategy since it reduces inventory waste by only manufacturing spare parts when they are required and by using less energy-intensive techniques. Siemens PGS's modular and upgradeable components, for example, are an example of this. Repairs may now be done on site, at the point of use, thanks to advances in AM repair technology. This is another advantage of the LMD AM technology that Royce and Rolls-Royce jointly developed. LMD may be used to repair damaged blisks and manufacture components in situ. As a result, the blisk's lifetime may be increased while its utilization can be maximized. Another piece of Pratt & Whitney-developed technology, LENS, has been commercialized by Optomec, a manufacturer of equipment.

Cold-spray AM has a long history in remanufacturing. Remanufacturing diesel engines has been Caterpillar's specialty since the 1970s, thanks to this technology. Since then, Caterpillar has developed and improved its Cat Reman remanufacturing operations, replacing products with a mix of new and old components before they broke. Product end-of-life returns from the previous 5 years have averaged 94%. High-grade remanufactured engines and components have the same quality as new ones, resulting in better profit margins. Remanufactured engines now include 40% new components, but this may be reduced to 25% by improving quality control, less scrapping of remanufactured parts (i.e. better availability of Reman components), more innovative repair processes, and the use of additive manufacturing. Innovative treatments for expensive components include metal-washing worn piston rod surfaces, followed by machining to restore them to like-new condition.

Additionally, this method may be used to repair engine heads and blocks that have been damaged by cracks or other issues. Caterpillar sees the benefits of remanufacturing from both a financial and environmental standpoint. However, remanufactured engines may cost up to 60% less than new ones, with Reman parts costing up to 40% less than new ones. Remanufacturing isn't only an advantage for Caterpillar. For manufacturers, remanufacturing has the potential to produce £5.6 billion in income and help create approximately 310,000 new manufacturing jobs in the United Kingdom alone, all while reducing greenhouse gas emissions (Lavery et al., 2013).

High-value component remanufacturing may be scaled up using hybrid technologies that mix additive and subtractive processes. For instance, Hybrid Manufacturing Technologies (HMT), a spin-out company, developed the AMBITTM multitask system in response to the promising results of the Innovate UK RECLAIM project. Allowing for entirely automated repair and reconfiguration, this cutting-edge hybrid technology addresses the automation problem head on. Remanufacturing high-volume, high-value components may now be done using the approach in a more efficient, accurate, and productive manner. As with LMD, turbine blades have been fixed using this technology.

8.6 AM AS A STEP TOWARDS SUSTAINABLE INDUSTRIAL SYSTEMS

Examples in this chapter highlight how AM is reshaping the industrial sector and enabling new types of sustainable production and consumption, as shown by the examples. The question posed at the outset of this chapter was: How may additive manufacturing allow more sustainable production and consumption models? These examples provide some tentative responses to that issue. A look at how AM is allowing for more sustainability, who is taking advantage of these possibilities, and how AM is altering the business model and manufacturing distribution follows.

8.6.1 AM Prospects for Facilitating Sustainability: How?

This chapter has mostly examined how AM may help sustainability from various aspects of a product life cycle perspective. At this stage of the product life cycle, AM is a novel manufacturing process that has immediate impact. As a result of this research, it can be shown that the product's life cycle is starting to profit from its environmental impact. Several early benefits are shown in this work, with the adoption and diffusion rates varying depending on the stage at which they were implemented. As a replacement for more conventional methods, additive manufacturing may also be used to produce customized small quantities or one-off objects. Increasing numbers of firms are embracing the technology or using service bureau solutions as a result of technical and commercial evidence of these advantages. For some companies, AM will be a direct replacement for their current manufacturing processes, while for others, it will serve as a complement or a vehicle to enter new markets. Second, AM offers a great deal of freedom in terms of design. It is possible to produce and save digital blueprints for additive manufacturing, enabling the fabrication of replacement parts on demand when repairs are needed. When modular design is paired with repair, remanufacturing, and refurbishment procedures, it is possible to extend and improve the lifespan of a product. The life cycle perspective of additive manufacturing and sustainable manufacturing on metal is suggested in a study (Peng et al., 2018), shown in Figure 8.7.

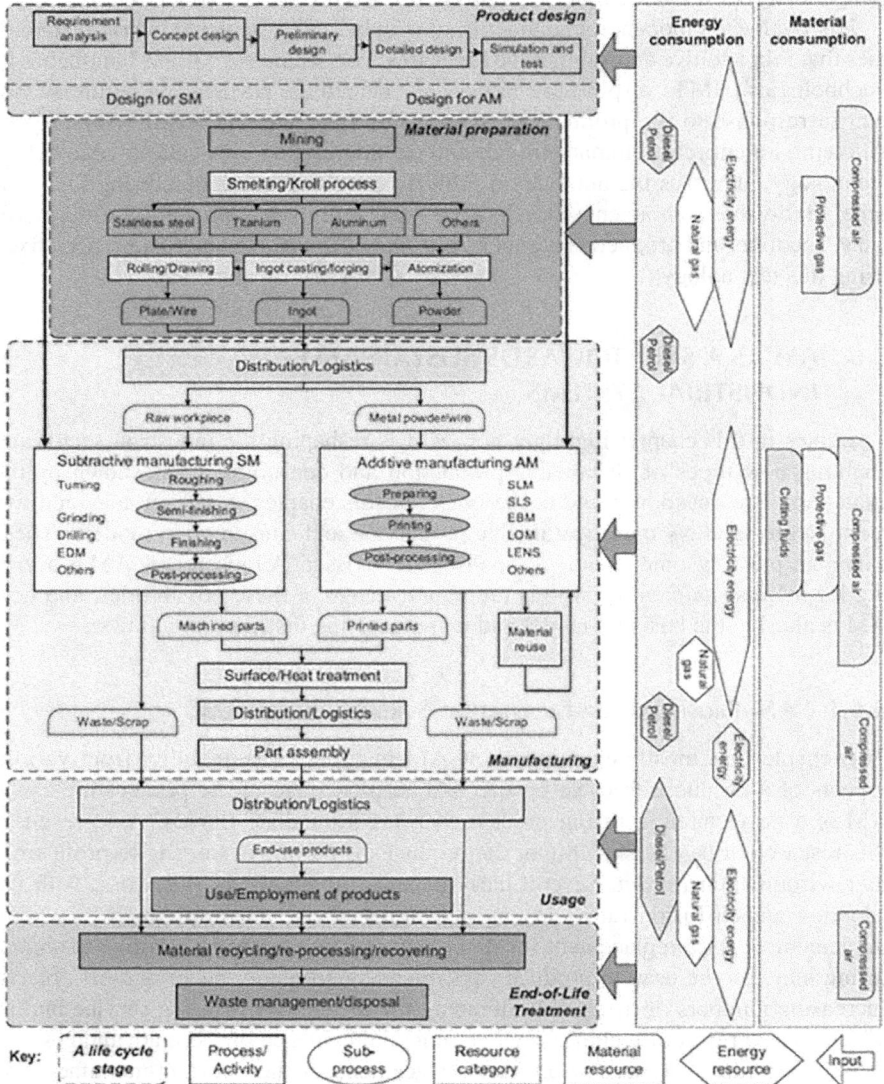

FIGURE 8.7 Life cycle perspective of additive manufacturing and sustainable manufacturing on metal (Peng et al., 2018).

8.6.2   Recognizing AM Prospects for Sustainability

Within this complex network of existing infrastructure, technologies, behaviors, customs, and attitudes are developed and accepted technology as a social construct (Bijker & Law, 1992; Metcalfe, 1998). People and organizations' investments in hard and soft technology drive complex systems to exhibit path dependence, as taught to us by the social construction of technology (David, 1985; Shapiro & Varian, 1999) In these complex systems, asynchronies between supply and demand provide opportunities for value development and capture (Ford et al., 2014; Metcalfe, 1998)

Environmental benefits are a welcome byproduct of the economic motivations of businesses competing in this market. There are instances when social or environmental values drive behavior, but most of the time, it is driven by the production or capture of economic value Filabot, for example, is a cleantech company dedicated to commercializing technology that has less harmful effects on the environment. Localized polymer recycling may benefit from Filabot's products, which take something that would otherwise be thrown away.

These tendencies are expected to continue in light of current market circumstances. To better serve present customers, well-established companies will rethink components and products and provide replacement parts as well as maintenance services. As long as economic and environmental objectives are kept in sync, this course of action will pay off in the long run. As a consequence, entrepreneurial businesses will experiment with new products and services in new markets, resulting in systemic change. Experimentation by large firms will be closely monitored. If niches are being seized by risky new firms, the latter might utilize a "wait and see" method before making a purchase of those that indicate growth potential or have already established a significant position (Christensen, 1997). Since the acquisition of Makerbot in 2013, Stratasys has used this technology.

8.6.3   AM and Sustainability in Business Models

Businesses use the business model to analyze how, for whom, and how they recoup the value they generate (Andries et al., 2013). Structured framework is used for the analysis that how a focused firm conducts business with customers, partners, and vendors. This encapsulates the firm's pattern of boundary-crossing relationships by considering the factor and product markets (Zott & Amit, 2007). AM adoption may necessitate a rethinking of existing business models, which could affect the long-term viability of those businesses. AM generates new business opportunities in the areas of repair, refurbishment, and remanufacturing. There is a growing understanding among businesses of how additive manufacturing can be used to extend product lifespans and close the loop. If AM technologies for repair, refurbishment, and remanufacturing are made available to businesses, it is hypothesized that product-service business models will become more common. Research is needed to better understand how AM-based manufacturing systems and supply/value chains may impact total resource consumption, given the lack of information currently available. Rather than focusing solely on the use of a single equipment, these should look at the larger production network and the life cycle analysis of components and products generated within networks.

8.6.4   RE-DISTRIBUTED MANUFACTURING

New business models and innovation opportunities arise as a result of AM being used in production. Since the Industrial Revolution, manufacturing has been more centralized. In contrast, modern digital manufacturing technologies like AM are enabling production to be decentralized. Previously dispersed industrial activity is being redistributed as more localized production becomes economically feasible (Pearson et al., 2014). The following is an example of an AM-based future vision:

> Future factories will be more diverse and scattered than those of today. It is expected that the manufacturing landscape would feature high-capital superfactories that produce complicated goods, as well as flexible units that can be reconfigured to meet the changing needs of their supply chain partners. As industries lessen their environmental effect, urban locations will become more widespread. As a result, the future factory may be at the patient's bedside or at home or at the workplace, or even on a battlefield.
>
> (BIS, 2013)

8.6.5   ADVANTAGES AND CHALLENGES IN SUSTAINABILITY WITH AM ADOPTION

There's a good chance that new applications of additive manufacturing with further environmental advantages will be developed soon. Product and material life cycles are used as a conceptualization model in this research to emphasize the environmental aspects of sustainability most, although sustainability comprises environmental, social, and economic aspects. Some components of social sustainability have evolved, but they are few and far between. This chapter does not include, for example, employment and the allocation of labor, health and safety, ethics, life satisfaction, and creative expression. Research on the social sustainability of AM must be expanded and supplemented, as is obvious from the current studies (Huang et al., 2013; Kohtala & Hyysalo, 2015).

8.7   DISCUSSION

There are several applications of additive manufacturing. Some key advantages of AM are: 3D printing is a frequent term for additive manufacturing; by building items layer by layer, additive manufacturing mimics biological processes; reducing the use of resources: additive manufacturing allows for improvements to be made in both the manufacturing and usage stages of the production process; there is less wastage of material; it also has a longer shelf life; and high-value seats can be modified to suit your needs. It is possible to have a more comprehensive knowledge of the implications of additive manufacturing for increasing industrial system sustainability by examining it in light of industrial sustainability.

In industry, by the use of additive manufacturing technologies, rapid prototyping is capable of producing printed components, not only models. However, rapid prototyping isn't always the best solution for every situation, and computer numeric control machine processes are sometimes necessary. In applied sciences like the automotive and aerospace industries, lightweight parts can be produced with the help of addictive manufacturing technologies. Additive manufacturing technologies have made it possible to manufacture complex cross-sectional areas like honeycomb cells or any other material part with cavities and cutouts, which reduces the weight-strength relationship. In AM printing technologies like selective laser sintering or electron beam melting, hollow structures should be attempted to manufacture less expensive products. Large-scale production is expected to become more cost-effective through additive manufacturing in the future. Improvements in strength, stiffness, energy efficiency, and corrosion resistance are all possibilities. Additive manufacturing has found particular application in the aerospace sector because of its performance requirements and low production scale. The manufacturing of aero-engine components has a significant environmental impact on the airline industry. Laser material deposition (LMD) used for the production of blade discs for aero-engines is one of the outputs of this enhanced AM method (Castro et al., 2015; Shamsaei et al., 2015). The advantage of this manufacturing technique is that it does not generate trash (swart). While recycling swart material is possible, it is generally done at a lower value stream and uses the same amount of energy as the original manufacturing process. There are a number of products that can only be made by using AM, such as tools and molds. The manufacturing process may be made more energy and resource saving.

It is necessary to automate current AM systems in order for production system integration to increase resource efficiency. AM technologies need to be developed further to allow for the incorporation of these additional features throughout the production process. Many benefits may be gained by using hybrid manufacturing techniques, including better end product quality, faster production times, and less tool wear. These benefits may also be reached via the combination of additive manufacturing with classic subtractive, joining, and transformational methods. With the construction industry being a large consumer of resources such as electricity and water, process redesign has the potential to have a considerable impact on the sector's resource efficiency. AM could change the value chain in the manufacturing industry. Simpler products with fewer parts, materials, actors, stages, and interactions could be the result of redesigning products and components. Product improvement through simplification can reduce supply chain scale and environmental impact throughout the entire supply chain. Transport emissions will be reduced while at the same time supporting and empowering local communities, as a more decentralized manufacturing system is implemented.

There are, however, obstacles and challenges that must be overcome before these benefits can be fully realized. In the first place, designers and engineers have an outdated idea of what additive manufacturing technology can do because its original application was restricted to rapid prototyping and tooling. Innovative shapes, such as microelectronics and other new functions, may be developed by AM. AM technologies require ongoing improvement to allow for the incorporation of additional functionality during production processes. AM technologies need to be rethought and improved in order to realize their full potential at the design stage.

Using AM as a starting point, there are chances to increase sustainability. A wide range of materials are employed in AM processes, resulting in a diverse range of products. Which AM method is chosen determines the kind of material that is used.

Material types include liquid, filament, powder, and solid sheets. The four primary concerns dealing with materials are:

1. Rethinking how specific raw materials are treated in order to minimize the resources required to get them into a useful condition as inputs for industrial processes is possible during the raw material processing stage.
2. Laser sintering and melting, as well as electron beam melting, are examples of the utilization of metal powders.
3. Retention and processing of metal ores in preparation for production use a significant amount of energy.
4. In the additive manufacturing process, recycled and virgin materials coexist as inputs. Simple 3D printers using FDM technology is a good fit. Few companies use these wood by-products and binding chemicals to generate a wood filament for AM. The recirculation of material is limited by quality and purity concerns, which may hinder recycling of items when they are no longer useful. Wood polymer-composite recycling is still in its infancy, and the necessary infrastructure is lacking in the United States.

The life cycle perspective of additive manufacturing and sustainable manufacturing for metal is suggested in a study (Peng et al., 2018), shown in Figure 8.7. Further investigation and validation of the mechanical as well as thermal characteristics of AM methods and materials are necessary in order to determine the most resource-efficient standards and allow this standardization.

8.8 SCOPE AND CONCLUSIONS

This overview presents how AM can facilitate more sustainable production and resource consumption strategies. It was found that the four main areas of AM's implementation from a life cycle perspective are product/process redesign, input material processing, make-to-order product manufacturing, and the closing the loop. It led to a better understanding of the advantages and disadvantages of AM for sustainability across product and material life cycles. To fully understand the advantages and disadvantages of this technology, a large amount of fresh study is required. At this exploratory level of innovation and ongoing research in AM and sustainability, such investigations include deep-dive single case studies in different sectors, organizations, products, and components, as well as models of AM-based production systems. An investigation of the ways in which AM impacts sustainability may help us get a more complete picture of how these advantages can be realized, as well as how and why they can't be realized. While this research focuses on the environmental impacts of AM, future studies should not disregard the social impacts of this unique manufacturing technique.

REFERENCES

Al-Lami, A., Hilmer, P., & Sinapius, M. (2018). Eco-efficiency assessment of manufacturing carbon fiber reinforced polymers (CFRP) in aerospace industry. *Aerospace Science and Technology*, 79, 669–678. https://doi.org/10.1016/j.ast.2018.06.020

Andries, P., Debackere, K., & Van Looy, B. (2013). Simultaneous experimentation as a learning strategy: business model development under uncertainty. *Strategic Entrepreneurship*, *7*(4), 288–310.

Baumers, M., Tuck, C., Bourell, D. L., Sreenivasan, R., & Hague, R. (2011). Sustainability of additive manufacturing: Measuring the energy consumption of the laser sintering process. *Proceedings of the Institution of Mechanical Engineers, Part B: Journal of Engineering Manufacture*, *225*(12), 2228–2239. https://doi.org/10.1177/0954405411406044

Bekas, D. G., Hou, Y., Liu, Y., & Panesar, A. (2019). 3D printing to enable multifunctionality in polymer-based composites: A review. *Composites Part B: Engineering*, *179*, 107540. https://doi.org/10.1016/j.compositesb.2019.107540

Bijker, W. E., & Law, J. (1992). *Shaping Technology/Building Society: Studies in Sociotechnical Change*. MIT Press.

BIS. (2013). *Foresight (2013): Future of Manufacturing: A New Era of Opportunity and Challenge for the UK Summary Report*. The Government Office for Science.

Braungart, M., & McDonough, W. (2002). *Cradle to Cradle: Remaking the Way We Make Things*. North Point Press.

Bryll, K., Piesowicz, E., Szymański, P., Slaczka, W., & Pijanowski, M. (2018). Polymer composite manufacturing by FDM 3D printing technology. *MATEC Web of Conferences*, *237*, 0–6. https://doi.org/10.1051/matecconf/201823702006

Castro, G., Rodríguez, J., Montealegre, M. A., Arias, J. L., Yañez, A., Panedas, S., & Rey, L. (2015). Laser additive manufacturing of high added value pieces. *Procedia Engineering*, *132*, 102–109. https://doi.org/10.1016/j.proeng.2015.12.485

Chen, D., Heyer, S., Ibbotson, S., Salonitis, K., Steingrímsson, J. G., & Thiede, S. (2015). Direct digital manufacturing: Definition, evolution, and sustainability implications. *Journal of Cleaner Production*, *107*, 615–625. https://doi.org/10.1016/j.jclepro.2015.05.009

Christensen, C. M. (1997). *The Innovator's Dilemma: When New Technologies Cause Great Firms to Fail*. Harvard Business School Press.

Daminabo, S. C., Goel, S., Grammatikos, S. A., Nezhad, H. Y., & Thakur, V. K. (2020). Fused deposition modeling-based additive manufacturing (3D printing): Techniques for polymer material systems. *Materials Today Chemistry*, *16*, 100248. https://doi.org/10.1016/j.mtchem.2020.100248

David, P. A. (1985). Clio and the economics of QWERTY. *American Economic Review*, *75*(2), 332–337.

de Jong, J. P. J., & de Bruijn, E. (2013). Innovation lessons from 3-D printing. *MIT Sloan Management Review*, *54*(2), 43–52.

Faludi, J., Bayley, C., Bhogal, S., & Iribarne, M. (2015). Comparing environmental impacts of additive manufacturing vs traditional machining via life-cycle assessment. *Rapid Prototyping Journal*, *21*(1), 14–33. https://doi.org/10.1108/RPJ-07-2013-0067

Ford, S. J., Routley, M. J., Phaal, R., & Probert, D. R. (2014). The industrial emergence of commercial inkjet printing. *European Journal of Innovation Management*, *17*(2), 126–143. https://doi.org/10.1108/EJIM-01-2013-0002

Fox, S., & Li, L. (2012). Expanding the scope of prosumption: A framework for analysing potential contributions from advances in materials technologies. *Technological Forecasting and Social Change*, *79*(4), 721–733. https://doi.org/10.1016/j.techfore.2011.10.006

Gartner, W. B. (1995). The describing framework for conceptual creation venture phenomenon of new venture creation. *Academy of Management Review*, *10*(4), 696–706. www.jstor.org/stable/10.2307/258039

Gebler, M., Schoot Uiterkamp, A. J. M., & Visser, C. (2014). A global sustainability perspective on 3D printing technologies. *Energy Policy*, *74*, 158–167. https://doi.org/10.1016/j.enpol.2014.08.033

Goh, G. D., Agarwala, S., Goh, G. L., Dikshit, V., Sing, S. L., & Yeong, W. Y. (2017). Additive manufacturing in unmanned aerial vehicles (UAVs): Challenges and potential. *Aerospace Science and Technology*, *63*, 140–151. https://doi.org/10.1016/j.ast.2016.12.019

Grossin, D., Montón, A., Navarrete-Segado, P., Özmen, E., Urruth, G., Maury, F., Maury, D., Frances, C., Tourbin, M., Lenormand, P., & Bertrand, G. (2021). A review of additive manufacturing of ceramics by powder bed selective laser processing (sintering/melting): Calcium phosphate, silicon carbide, zirconia, alumina, and their composites. *Open Ceramics*, *5*(January). https://doi.org/10.1016/j.oceram.2021.100073

Guo, N., & Leu, M. C. (2013). Additive manufacturing: Technology, applications and research needs. *Frontiers of Mechanical Engineering*, *8*(3), 215–243. https://doi.org/10.1007/s11465-013-0248-8

Gutowski, T. G., Branham, M. S., Dahmus, J. B., Jones, A. J., Thiriez, A., & Sekulic, D. P. (2009). Thermodynamic analysis of processes. *Environmental Science & Technology*, *43*(5), 1584–1590.

Hao, L., Raymond, D., Strano, G., & Dadbakhsh, S. (2010). Enhancing the sustainability of additive manufacturing. *IET Conference Publications*, *2010*(565 CP), 390–395. https://doi.org/10.1049/cp.2010.0462

Huang, S. H., Liu, P., Mokasdar, A., & Hou, L. (2013). Additive manufacturing and its societal impact: A literature review. *International Journal of Advanced Manufacturing Technology*, *67*(5–8), 1191–1203. https://doi.org/10.1007/s00170-012-4558-5

Kohtala, C., & Hyysalo, S. (2015). Anticipated environmental sustainability of personal fabrication. *Journal of Cleaner Production*, *99*, 333–344. https://doi.org/10.1016/j.jclepro.2015.02.093

Lavery, G., Pennell, N., Brown, S., & Evans, S. (2013). *Next Manufacturing Revolution Available*. www.nextmanufacturingrevolution.org/nmr-reportdownload/

Liu, J., Sun, L., Xu, W., Wang, Q., Yu, S., & Sun, J. (2019). Current advances and future perspectives of 3D printing natural-derived biopolymers. *Carbohydrate Polymers*, *207*, 297–316. https://doi.org/10.1016/j.carbpol.2018.11.077

Lubik, S., & Garnsey, E. (2016). Early business model evolution in science-based ventures: The case of advanced materials. *Long Range Planning*, *49*(3), 393–408. https://doi.org/10.1016/j.lrp.2015.03.001

Mani, M., Lyons, K. W., & Gupta, S. K. (2014). Sustainability characterization for additive manufacturing. *Journal of Research of the National Institute of Standards and Technology*, *119*, 419–428. https://doi.org/10.6028/jres.119.016

Mellor, I., Grainger, L., Rao, K., Deane, J., Conti, M., Doughty, G., & Vaughan, D. (2015). Titanium powder production via the metalysis process. In *Titanium Powder Metallurgy: Science, Technology and Applications*. Elsevier Inc. https://doi.org/10.1016/B978-0-12-800054-0.00004-6

Metcalfe, J. S. (1998). *Evolutionary Economics and Creative Destruction*. Routledge.

Neely, A. (2008). Exploring the financial consequences of the servitization of manufacturing. *Operations Management Research*, *1*(2), 103–118. https://doi.org/10.1007/s12063-009-0015-5

Pearson, H., Noble, G., & Hawkins, J. (2014). *Re-distributed Manufacturing Workshop Report*, 7e8 November. https://www.epsrc.ac.uk/newsevents/pubs/re-distributed-manufacturing-workshop-report/ (accessed 04.02.2022).

Peng, T., Kellens, K., Tang, R., Chen, C., & Chen, G. (2018). Sustainability of additive manufacturing: An overview on its energy demand and environmental impact. *Additive Manufacturing*, *21*(April), 694–704. https://doi.org/10.1016/j.addma.2018.04.022

Petrick, I. J., & Simpson, T. W. (2013). 3D printing disrupts manufacturing. *Research Technology Management*, *56*(6), 12–16. https://doi.org/10.5437/08956308X5606193

Petrovic, V., Vicente Haro Gonzalez, J., Jordá Ferrando, O., Delgado Gordillo, J., Ramon Blasco Puchades, J., & Portoles Grinan, L. (2011). Additive layered manufacturing: Sectors of industrial application shown through case studies. *International Journal of Production Research, 49*(4), 1061–1079. https://doi.org/10.1080/00207540903479786

Phaal, R., O'Sullivan, E., Routley, M. J., Ford, S. J., & Probert, D. R. (2011). A framework for mapping industrial emergence. *Technol. Forecast. Soc. Change, 78*(2), 217–230.

Sandstrom, C. (2015). Adopting 3D printing for manufacturing—evidence from the hearing aid industry. *Technological Forecasting and Social Change, 102*, 160–168.

Shamsaei, N., Yadollahi, A., Bian, L., & Thompson, S. M. (2015). An overview of direct laser deposition for additive manufacturing; Part II: Mechanical behavior, process parameter optimization and control. *Additive Manufacturing, 8*, 12–35. https://doi.org/10.1016/j.addma.2015.07.002

Shapiro, C., & Varian, H. R. (1999). *Information Rules: A Strategic Guide to the Network Economy*. Harvard Business School Press.

Silva, M., Felismina, R., Mateus, A., Parreira, P., & Malça, C. (2017). Application of a hybrid additive manufacturing methodology to produce a metal/polymer customized dental implant. *Procedia Manufacturing, 12*(December 2016), 150–155. https://doi.org/10.1016/j.promfg.2017.08.019

Sreenivasan, R., Goel, A., & Bourell, D. L. (2010). Sustainability issues in laser-based additive manufacturing. *Physics Procedia, 5*(PART 1), 81–90. https://doi.org/10.1016/j.phpro.2010.08.124

Teuber, L., Osburg, V. S., Toporowski, W., Militz, H., & Krause, A. (2016). Wood polymer composites and their contribution to cascading utilisation. *Journal of Cleaner Production, 110*, 9–15. https://doi.org/10.1016/j.jclepro.2015.04.009

TSB. (2012). *Shaping Our National Competency in Additive Manufacturing: Technology Innovation Needs Analysis Conducted by the Additive Manufacturing Special Interest Group for the Technology Strategy Board, Technology Strategy Board Knowledge Transfer Network Special*. Reeves Insight The Silversmiths.

Yin, R. K. (2009). *Case Study Research: Design and Methods*. Sage.

Zhou, F., Ji, Y., & Jiao, R. J. (2013). Affective and cognitive design for mass personalization: Status and prospect. *Journal of Intelligent Manufacturing, 24*(5), 1047–1069. https://doi.org/10.1007/s10845-012-0673-2

Zott, C., & Amit, R. (2007). Business model design and the performance of entrepreneurial firms. *Organization Science, 18*(2), 181–199. https://doi.org/10.1287/orsc.1060.0232

9 A Comprehensive Review of Design and Development of Advanced Tailored Material on Sustainability Aspects

Saurabh Rai, Subodh Kumar, Ankit Gupta

CONTENTS

DOI: 10.1201/9781003269298-9

9.1 INTRODUCTION

The advancement of technology has given rise to new materials that meet sophisticated requirements. The need for advanced material has led to advanced high-strength alloy composites that can push limitations, such as in the automobile, aerospace, and biomedical sectors (Egbo 2021). In this context, several categories of advanced composites have been explored. Figure 9.1(a) shows that a composite consists of a matrix and fibers with specific orientations to give the required properties. Fibers are materials that provide strength, and the matrix adheres to the fiber and correctly places it. Composites provide superior properties like stiffness-weight ratio and strength-weight ratio with dimensional stability and design flexibility. Their properties are making composites rapidly replace conventional material, especially in the aerospace and automobile industries (Sharma, Bhandari, and Ashirvad 2014). The problem with conventional composite materials are significant stress variation leading to delamination, stress concentration matrix necking, and adhesive bond separation (Ribeiro 2018). These problems of multilayer composites can be resolved using graded material.

Functionally graded materials (FGMs) are advanced inhomogeneous high-performance composite materials with tailored properties for a specific application. FGMs have continuous gradation in properties from one surface to another (Ghatage, Kar, and Sudhagar 2020; Jha, Kant, and Singh 2013; Niknam et al. 2018). FGMs have composition variations without any sharp interface boundary between different compositions. The concept of the development of FGMs started from 1972. However, FGMs can be found in nature from a million years ago. In plants and animal tissues, teeth are good examples of FGMs with properties varying gradually for specific functions. Property variation in FGM is accomplished by changing the composition of the material. Figure 9.1(b) shows a material variation from top to bottom. It is called step graded material. The properties remain the same for one layer; these properties change as the height varies. Figure 9.1(c) shows the proper gradation of FGM. This type of FGM results in an extraordinarily smooth variation of the properties.

FGM was initially conceptualized by Bever and Duwez (1972). The researchers studied the theoretical aspect of graded structured composites, and the first FGM

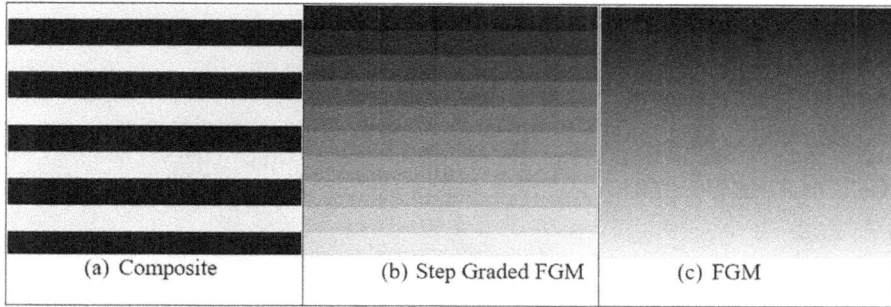

FIGURE 9.1   Different advanced materials.

was developed and used by Japanese scientists in 1984 as a thermal barrier material (Koizumi 1997). The FGM was developed to minimize thermal stress using a metal-ceramic FGM in a reusable rocket engine. FGM is successfully used in heat exchangers to have conductivity on the inner surface and isolation on the outer surface (Burlayenko et al., 2017). An FGM of Al-Al$_2$O$_3$ combines the benefits of metals and ceramics in a single material (Karandikar et al. 2013). Several reviews on various aspects of FGM have been published in recent years (Gupta and Talha 2015a; Kumar Bohidar, Sharma, and Mishra 2014; Thai and Kim 2015). The latest research on FGMs and their manufacturing, mechanics, thermal characteristics, and applications is given in this chapter. Gupta and Talha (2015) presented a thorough examination of FGM structures, production techniques and several new applications in aerospace and healthcare (Khorsand, Fu, and Tang 2019).

In this chapter, the second section covers homogenization and gradation on rules that most researchers follow to evaluate FGM properties. These properties are used to calculate the effective properties of FGM. The third section discusses different manufacturing processes such as chemical vapor deposition (CVD), physical vapor deposition (PVD), plasma spraying, and ion mixing (Saleh et al. 2020). In the fourth section, the sustainability and feasibility aspects of different FGM manufacturing processes are discussed.

9.2   GRADATION AND HOMOGENIZATION RULE

Analysis of FGM structures requires a proper model that can efficiently predict the material's effective properties. There is a continuous change of properties in FGM, making it very difficult to evaluate them. Many factors like shape, size, and distribution phase play a substantial role in predicting the actual properties. Overall properties like the Young's modulus and Poisson's ratio of FGM plates are computed using different techniques, called homogenization. Several models that can effectively evaluate the macroscopically homogeneous properties of FGM are discussed in the literature. Some standard techniques used for calculation of FGM properties are discussed in the following.

9.2.1 Voight Estimate

The Voight estimate method calculates properties of FGMs that have a uniform distribution of material without porosity. The material properties can be calculated using the rule of the mixture. In this material model, the properties at any location can be calculated by multiplying the properties corresponding to the volume fraction available at that layer (Gupta and Talha 2015).

$$P = V_\alpha P_\alpha + V_\beta P_\beta \tag{9.1}$$

Equation 9.1 is used to calculate unified properties like Young's modulus, density, and Poisson's ratio of FGM at any place with different material constituents.

9.2.2 Reuss Estimate

This relation is developed by assuming uniform stress distribution through the material. Material properties are calculated using the rule of mixture. The Reuss model expression gives the mean harmonic properties (Toudehdehghan et al. 2017).

$$P = \frac{P_\alpha P_\beta}{V_\alpha P_\beta + V_\beta P_\alpha} \tag{9.2}$$

9.2.3 Mori-Tanaka Scheme

The Mori-Tanaka scheme is used to predict the effective properties of FGM with a discontinuous material phase. The material's Young's modulus and shear modulus can be evaluated using Equations 9.3 to 9.9. These equations consider inclusion and its interaction with other constituents.

$$\frac{K_Z - K_m}{K_c - K_m} = \frac{V_f^P}{1 + (1 - V_f^P)\left(\dfrac{K_c - K_m}{K_m + \dfrac{4}{3}G_m}\right)} \tag{9.3}$$

In this, G is shear modulus, K represent the bulk modulus, V_p represent volume fraction, and z is any height.

$$\frac{G_Z - G_m}{G_c - G_m} = \frac{V_f^P}{1 + (1 - V_f^P)\left(\dfrac{G_c - G_m}{G_m + \dfrac{4}{3}f_m}\right)} \tag{9.4}$$

$$f_m = \frac{G_m(9K_m + 8G_m)}{6(K_m + 2G_m)} \qquad (9.5)$$

The G and K are calculated from Eqs. (3) and (4) to calculate the Young's modulus and Poisson's ratio.

$$E_m = \frac{9K_m G_Z}{3K_Z + G_Z} \qquad (9.6)$$

$$v = \frac{3K_z - 2G_Z}{2(3K_Z + G_Z)} \qquad (9.7)$$

Calculation of the thermal properties can be done similarly to the shear calculation. Eq. (9.8) shows the analysis of heat conductivity. In the equation, K represent thermal conductivity and thermal expression with all subscripts having the same meaning as previously (Benveniste 1987; Mori and Tanaka 1973; Tanaka et al. 1993).

$$\frac{K_Z - K_m}{K_c - K_m} = \frac{V_f^P}{1 + (1 - V_f^P)\left(\dfrac{K_c - K_m}{3K_m}\right)} \qquad (9.8)$$

$$\frac{\alpha_z - \alpha_m}{\alpha_c - \alpha_m} = \frac{\dfrac{1}{K_z} - \dfrac{1}{K_m}}{\dfrac{1}{K_c} - \dfrac{1}{K_m}} \qquad (9.9)$$

9.2.4 GRADATION RULES

Gradation rules are used to calculate the properties of the FGM at a given value of height (z). The effective mechanical and thermal properties varies along the thickness in unidirectional FGM structures can be estimated using power-law, exponential, and sigmoidal-law distribution as given below:

9.2.5 POWER-LAW

The material properties vary in "z" direction can be calculated using power-law distribution (Zenkour 2006) as given in Eq. (9.10).

$$P(z) = P_m + (P_c - P_m)V_c(z) \qquad (9.10)$$

P(z) denotes the effective material's property at height z, while subscripts c and m signify ceramic and metal, respectively. For unidirectional FGM, the material characteristics of FGMs are determined by the volume fraction V_c, which follows the power-law provided by Eq. 9.2.

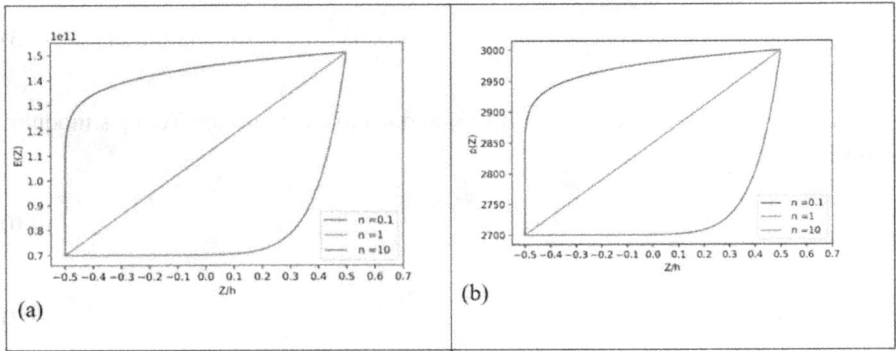

FIGURE 9.2 (a) Young's modulus; (b) density variation of the Al-zirconia FGM.

$$V_c = \left(\frac{z}{h} - \frac{1}{2}\right)^n \qquad (9.11)$$

In Figure 9.2, a typical variation for the properties of Al-zirconia using power-law along the transverse direction is depicted. The Young's modulus of the Al is 70 GPa, and the Young's modulus of the Zirconia is 150 GPa; this leads to a variation of the properties from 70 GPa (Metal) with at $-0.5h$ ratio to 150 GPa (ceramic) properties at $0.5h$. The volume fraction index n governs the gradation of mechanical properties between the top and bottom surface of the FGM plate The higher value of n depicts metal rich behavior, and the lower value of n denotes ceramic dominance. The trend variation of the density is the same as the Young's modulus and is shown in Figure 9.2(b).

9.2.6 Exponential Law

The material properties vary exponentially in one direction when the exponential law distribution is used for unidirectional FGMs, as expressed in Eq. (9.12). In general, the exponential law is employed to study the fracture and fatique behavior of FGM (El Harti et al. 2019) structures

$$P(z) = P_m e^{\left(\frac{1}{h}\right)\left(\ln\frac{P_c}{P_m}\right)\left(z+\frac{h}{2}\right)} \qquad (9.12)$$

P_z is the material properties in the z-direction as depicted in Figure 9.3; P_c and P_m have their usual meaning as the bottom and top surface properties, respectively.

9.2.7 Sigmoid Law Distribution

The exponential and power-law function gives the maximum stress at one of the interfaces. This can lead to stress concentration and material failure. In sigmoid law,

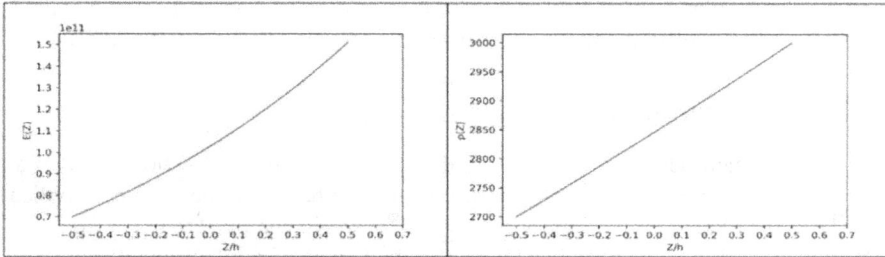

FIGURE 9.3   (a) Young's modulus; (b) density variation of the Al-Zirconia FGM using exponential law.

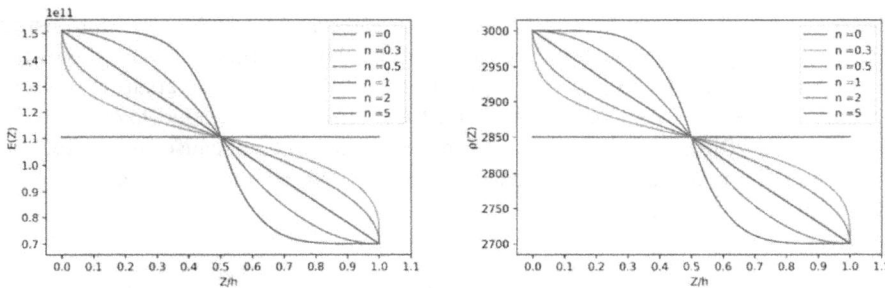

FIGURE 9.4   (a) Variation for the Young's modulus; (b) density variation with Z/h ratio in the sigmoid law.

the stress concentration can be circumvented by using two different power laws (Jung and Han 2013; Mathews and Murthy 2017). In this, as shown in Figure 9.4, there is gradation in two directions, which gives a smooth transition of the property and makes it a better gradation technique

The material properties are varied in one direction when considering the sigmoid law distribution. This law, applicable for unidirectional FGM, is given by Eqs. (9.13) and Eqs. (9.14) (Pascon and Coda 2013). The variation of Young's modulus and density using sigmoid law is shown in Figure 9.4.

$$P(z) = P_m \left[\frac{1}{2} \left(\frac{2z}{h} \right)^k \right] + P_c \left[1 - \frac{1}{2} \left(\frac{2z}{h} \right)^k \right] \quad 0 < z < \frac{1}{2}h \qquad (9.13)$$

$$P(z) = P_m \left[1 - \frac{1}{2} \left(\frac{2-2z}{h} \right)^k \right] + P_c \left[\frac{1}{2} \left(\frac{2-2z}{h} \right)^k \right] \quad \frac{1}{2}h < z < h \qquad (9.14)$$

These are some material homogenization techniques used to analyze FGM plates. Sometimes these properties are integrated into the thickness to calculate comparable properties of the isotropic plate (Kumar and Jana 2019). This simple material model converts the properties of FGM like orthotropic material, making it easy to analyze the strength, natural frequency, and other parameters (Singh and Gupta 2022).

9.3 FABRICATION TECHNIQUES OF FGM

The different methods for producing FGM are discussed in this section. FGM is an advanced composite material that can be customized based on the required applications (El-Galy, Saleh, and Ahmed 2019a). As shown in Figure 9.5, FGM manufacturing can broadly be classified according to the nature of production (Saleh, Jiang, Fathi et al. 2020),.

9.3.1 DEPOSITION METHOD

Generally, FGM is manufactured by depositing the material in a substrate using a thermal spray, with electrophoretic deposition. This method is widely used for making microscale FGM coating applications. This method is versatile and can be implemented to make one-, two-, and three-directional graded layers (Deng, Kim, and Ki 2019). The proper gradation can be achieved because the deposition evaporates the material This process is slow and energy intensive and can make many parts.

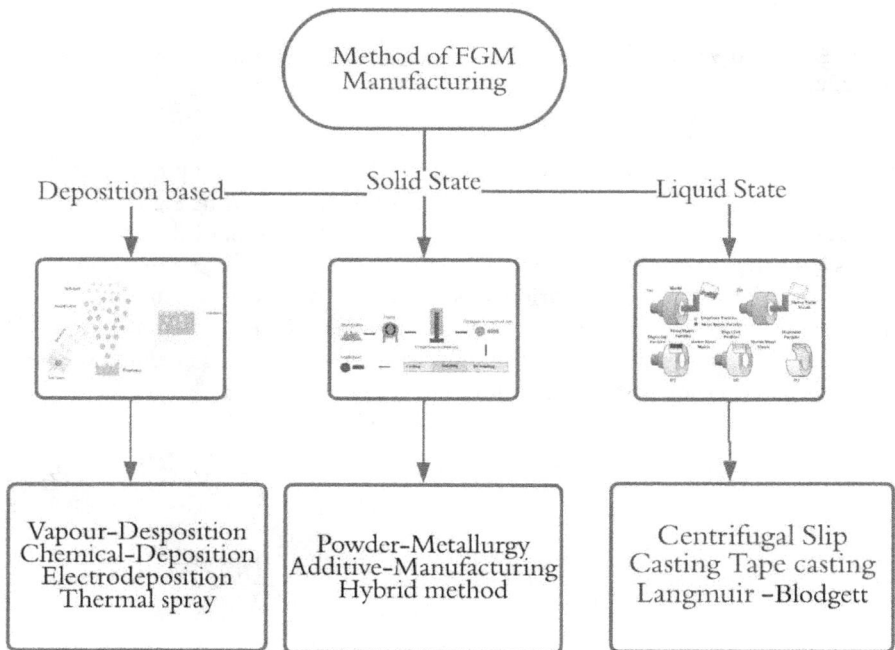

FIGURE 9.5 Classification of manufacturing methods for FGM and fabrication processes (Łatka et al. 2020).

9.3.2 Vapor Deposition

The material is manufactured by depositing a thin layer of graded material in this process. The layers are generally ion the order of nm to sub-μm. The vapor is deposited and condensed in this process to get the FGM (Depeursinge et al. 2010). Initially, this method was used to produce coatings to provide additional material properties like mechanical properties, wear corrosion, and thermal properties on the surface. This deposition comprises two groups based on the formation of the vapor (Gupta and Talha 2015a)

9.3.3 Ion Beam–Assisted Deposition

In ion beam–assisted deposition (IBAD) in high vacuum settings, thin film deposition is achieved by combining evaporation with simultaneous ion beam bombardment, as shown in Figure 9.6. This approach differs from previous deposit procedures due to using an ion beam. High-velocity particles in various morphologies are used to create films, which is crucial to thin coatings (Popoola et al. 2016).

9.3.4 Physical Vapor Deposition (PVD)

In this method, gradation is produced by producing the material in vapor form and then condensing it back to form a film. The composition of the material can be changed by changing the rate of the vapor production of a different material. Since the material is produced in vapor form, it has a very uniform gradation and can be employed for metals and alloys. The quality of this process is significantly superior to other manufacturing techniques (Rajak et al. 2020). PVD uses electron beam deposition, evaporative deposition, cathodic arc deposition, and the sublimation sandwich method (Reinhold et al. 2004).

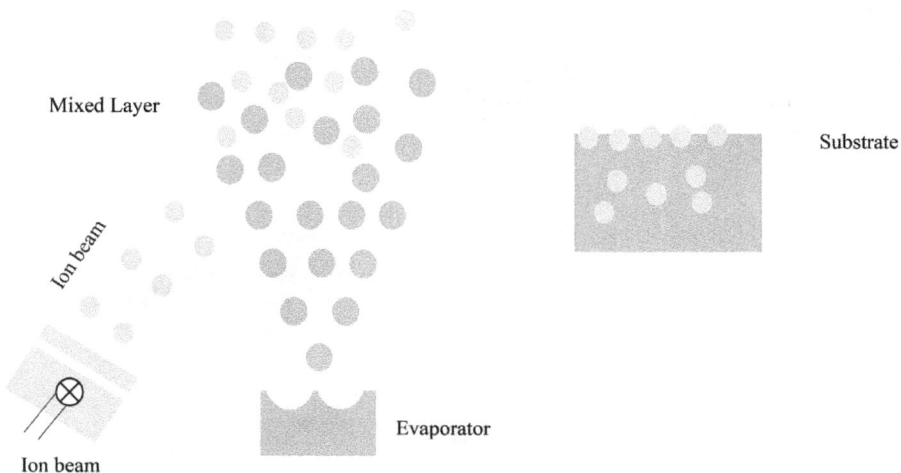

FIGURE 9.6 Ion beam-assisted deposition (Popoola et al. 2016).

9.3.5 CHEMICAL DEPOSITION

Chemical vapor deposition (CVD) is a technique that can be utilized to manufacture good-quality FGM. In the CVD process, a thin film of the target can be created by atmospheric pressure, low pressure, hot filament, electron-assisted, laser-assisted, and direct liquid injection.

The substrate is inserted in the chamber, as shown in Figure 9.7. The substance is coated with target material inside the vacuum by vaporizing the coating material (Choy 2003). This technique is very efficient for applying wear coating.

9.3.6 ELECTRODEPOSITION

Electrophoretic deposition (EPD) is a material-production technology based on the concept of electrolysis, or a chemical reaction. In this process, different particles go for the reduction reaction; these reduced particles produce the coating, and the material is placed on the substrate (Kaya et al. 2003).

This procedure employs contact with the solvent and additives to charge the particles. The charged particles are then passed under the control of the applied electrical field. As shown in Figure 9.8, the particles eventually form an enlarged accumulation of the deposition on electrode. This procedure can easily be employed to produce FGM.

9.3.7 THERMAL SPRAY METHOD

The thermal spray method can be employed for FGM to form a thin-dimensional surface coating through the spraying process. The components are protected by the surface graded layer because their nature necessitates suitable thermal, corrosion, wear, and electrical isolation (Heimann 2007). Thermal spray techniques are widely utilized to produce FGM, reported in the literature (Zhang et al. 2019). As shown in

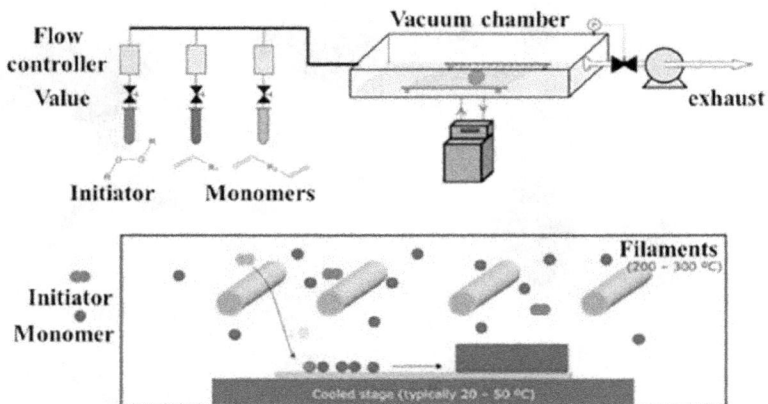

FIGURE 9.7 Schematic diagram of the CVD process (Saleh, Jiang, Fathi et al. 2020). {permission has been taken}

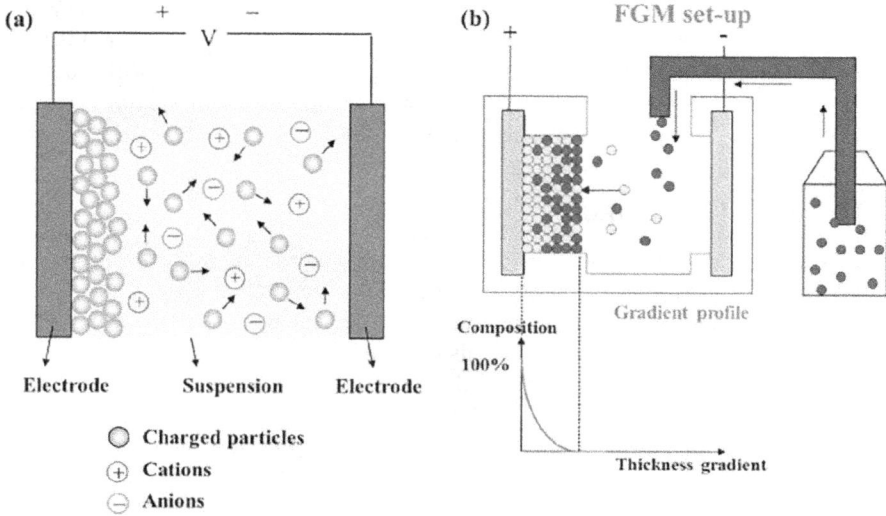

FIGURE 9.8 (a) EPD process; (b) EPD process for FGM (Saleh, Jiang, Fathi et al. 2020). {permission has been taken}

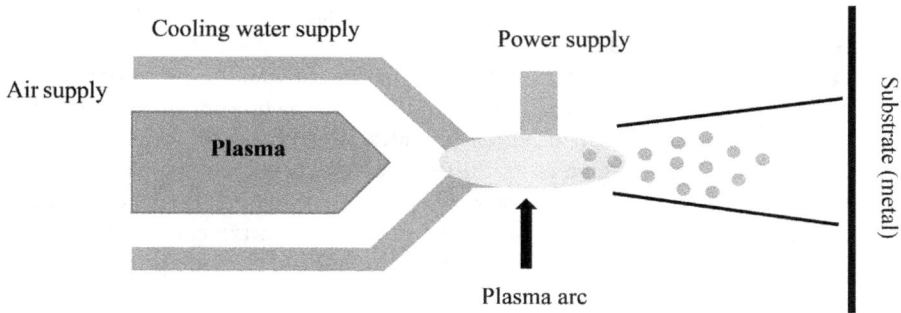

FIGURE 9.9 Schematic diagram of the thermal spray process (Zhang et al. 2019). {permission has been taken}

Figure 9.9, plasma is used to vaporize the target material that can be deposited on the substrate.

9.3.8 PLASMA SPRAYING

Plasma spraying is a thermal spraying procedure that melts and accelerates small particles onto a prepared surface using a high-energy heat source (Łatka et al. 2020). These molten particles (droplets) cool and solidify quickly upon impact due to heat transmission to the underlying substrate, forming a lamellae-like covering through accumulation. Flame spraying, plasma spraying, high-velocity oxygen fuel (HVOF), vacuum plasma spraying, arc metallization, and detonation gun spraying are the most common thermal spraying procedures. The process has been successfully used to manufacture bioactive bioceramic coatings on metal substrates, and flame spraying, HVOF spraying, and plasma spraying (Heimann 2007).

9.3.9 SOLID-STATE METHOD

FGM is produced in solid state using friction stir additive manufacturing, powder metallurgy, and additive manufacturing. The solid-state method is the most efficient process to make FGMs in a large size. These methods can be applied to mass production and controlled to high precision. This method is widely used in industrialization applications (El-Galy, Saleh, and Ahmed 2019b)

9.3.10 POWDER METALLURGY

Powder metallurgy (PM) is a process that produces functionally graded material (Nejad, Alamzadeh, and Hadi 2018; Sobczak and Drenchev 2013) using powders of different materials. This procedure involves weighing and combining powders in a pre-determined spatial distribution according to the functional requirements, ramming the premixed powders, and finally sintering. The sintering process is depicted in Figure 9.10. A progressive structure is created using the PM approach. PM is one of the most common methods, as it is simple and can apply to FGM mass production (Gasik, Kawasaki, and Ueda 2006).

In the stacking process, there should be spatial distribution of the materials with height. The compressing or compacting step mold made by powder with the required composition of the powder distribution provides strength to the mold for further processing of the part (Saxena et al. 2019). The compacted part is then taken for the sintering process. The sintering process is used to create bonds between materials; the packed powder is heated below the melting point, typically two-thirds to three-fourths of the melting point of powder material. Hot isostatic pressing (HIP),

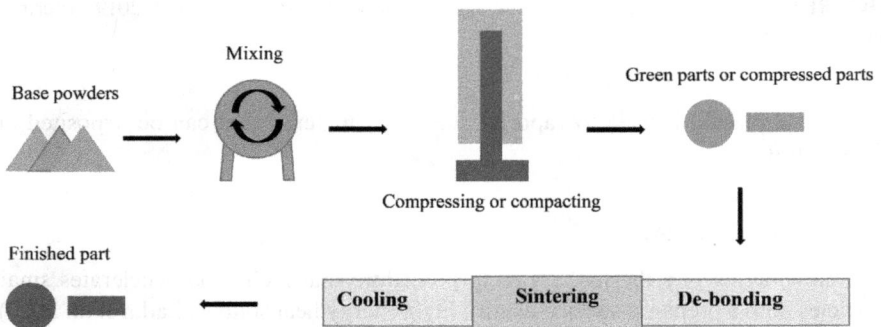

FIGURE 9.10 Powder metallurgy (Heimann 2007).

pressureless sintering, hot press, and spark plasma sintering are all options for sintering (SPS) (Heimann 2007).

9.3.11 Friction Stir Process

Friction stir is a solid-state processing technique that is very similar to friction stir
welding and is becoming a well-known technique for processing material (El-Galy,
Saleh, and Ahmed 2019b). The experimental setup for friction stir is shown in Figure
9.11. A milling machine is used to rotate and feed the tool; there is a fixture to hold
the sheet correctly, and the tool has a pin that gets inside the material. This process
tool, which is non-consumable and generally made of rigid material (high carbon
steel), has a pin on its tip that penetrates in the sheet. The tool is rotated and traversed
in the forward direction, as shown in Figure 9.11 (Singh et al. 2020) This localized
heat is generated due to internal deformation energy and interfacial friction between
tool and workpiece. An Al/SiC surface composite using the FSP technique was found
to have a uniform distribution of SiC in the Al matrix (Gandra 2010).

9.3.12 Additive Manufacturing

The development of the additive manufacturing (AM) process has led to significant
growth of FGM production. The material is manufactured layer by layer, which
allows much room to change the composition of the material. Some commonly available AM processes are laser-based methods, stereolithography, and material jetting
(Zhang et al. 2019).

9.3.13 Selective Laser Sintering

Selective laser sintering (SLS) uses laser energy to produce 3D parts of the desired
shape by stacking them layer by layer. The 3D geometry is sliced to a minimal thickness, and the product is manufactured by joining these layers. Figure 9.12 show a
schematic diagram of the SLS process. The powder layer is produced; then, the laser
is illuminated to get the required object on the desired area. The illuminated part is
bound, and the rest powder can be reused (Maskery et al. 2016).

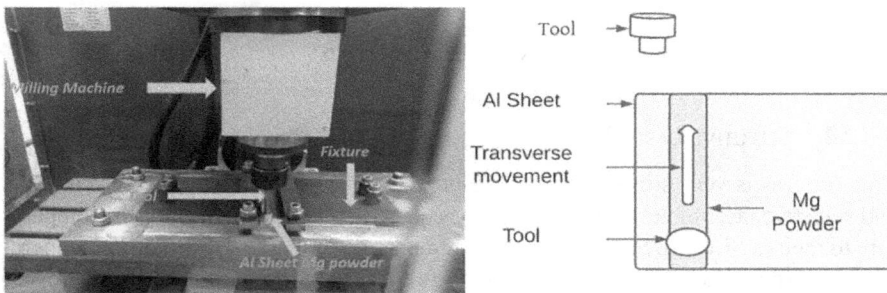

FIGURE 9.11 (a) FSW setup; (b) schematic of FSP process.

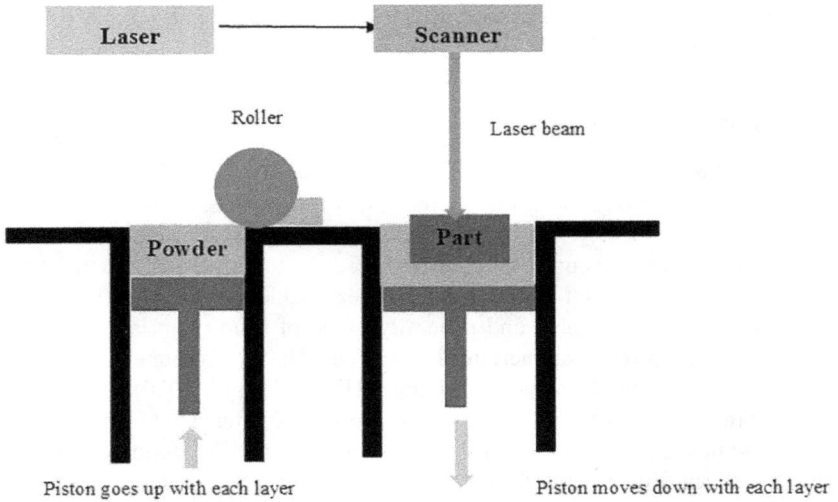

FIGURE 9.12   Selective laser sintering (Rider et al. 2018).

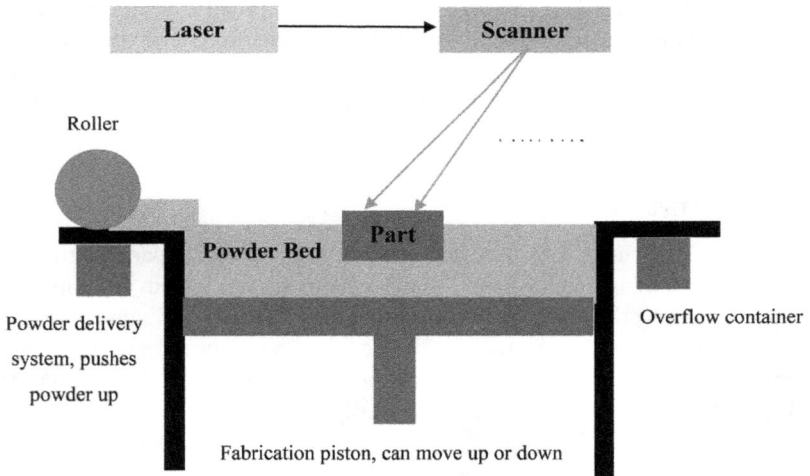

FIGURE 9.13   Selective laser melting (Jiao et al. 2018).

9.3.14   SELECTIVE LASER MELTING

This process is very similar to the SLS process. The only difference between SLS and selective laser melting (SLM) is that the layer is bounded by raising the temperature to melt in SLM. In SLS, the working temperature is 85% of the melting temperature. Figure 9.13 shows the proper working of the SLM process. The SLM process fabricates a cobalt base alloy. It was found that the material had a 48% and 60% more flexible FGM femoral stem than standard (Hazlehurst, Wang, and Stanford 2014).

9.3.15 DIRECT ENERGY DEPOSITION

In this process, the material is deposited using a power source. The feed rate of the material is adjusted to attain proper gradation for the FGM. As shown in Figure 9.14, the material comes from a powder feeder that contains different powder materials. This powder comes out through the nozzle; the laser is illuminated on the material to the desired temperature for bonding between the layers (Yan, Battiato, and Fadel 2018). A laser, arc, or electron- beam can be the energy source to bind the materials (Das et al. 2013).

9.3.16 HYBRID MANUFACTURING

Generally, a hybrid process uses a mix of two technologies to produce the material. Some processes like AM are very costly to take into mass production. This leads to the development of a hybrid process which combine AM with some conventional process to possess properties of AM while also being cost effective (Pan et al. 2018).

9.3.17 WIRE AND ARC MANUFACTURING

Arc welding is used with the AM process. This makes the production of FGM cheap and fast (Rodrigues et al. 2019). There have been the development of production of FGM using AM and plasma arc-welding (PAW), tungsten inert arc welding, and metal inert gas (MIG) plasma transferred arc (PTA) welding (Reisgen, Sharma, and Oster 2019). The schematic of the process is shown in Figure 9.15 the energy produced by the TIG torch is used to fuse the wire or powder. Once the layer is formed, the table has the other material layer.

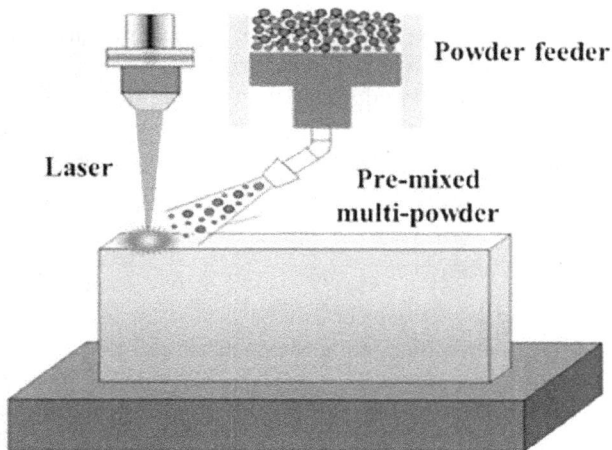

FIGURE 9.14 Laser-based method for the production of the FGM with DED (Saleh, Jiang, Fathi et al. 2020). {permission has been taken}

(a)   **(b)**

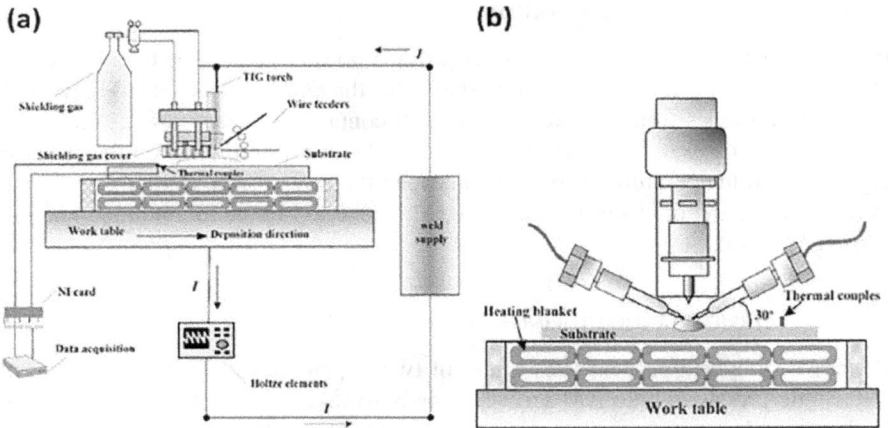

FIGURE 9.15 (a) Concept of wire and arc manufacturing process for producing FGM; (b) double-wire feeding units (Wang et al. 2018). {permission has been taken}

9.3.18 LIQUID STATE METHOD

In this method, the material is taken to the liquid state; then a process is performed to get gradation of the structure. This technique can produce a proper gradation of the material with less cost. However, there are significant challenges on controlling grading and wettability in the molten state (Chao et al. 2019; Saleh, Jiang, Ma et al. 2020).

9.3.19 CENTRIFUGAL FORCE METHOD

This is one of the most commonly used processes for manufacturing FGM for industrial applications (Watanabe, Eryu, and Matsuura 2001).

In this process, gradation is attained by the uneven force generated by the centrifugal force, as shown in Figure 9.16. Generally, the value of the force is high at a higher radius compared to a lower radius; this creates an uneven distribution of two or more constituents in the composite material.

9.3.20 SLIP CASTING METHOD

In this process, a suspension of fine ceramic particles is made on liquid metals. The process is shown in Figure 9.17. The Ni (one constituent) particles are kept on the steel surface, and molten Al (second constituent) is applied after them. As centrifugal force is applied, the particles are dispersed in the molten Al, forming a graded structure. The required thickness of the FGM is kept, and the remaining molten Al fluid is removed. The dry cast is removed from the mold. The advantage of slip casting is to build complicated structures and make them continuously graded for the production of FGMs (Depeursinge et al. 2010).

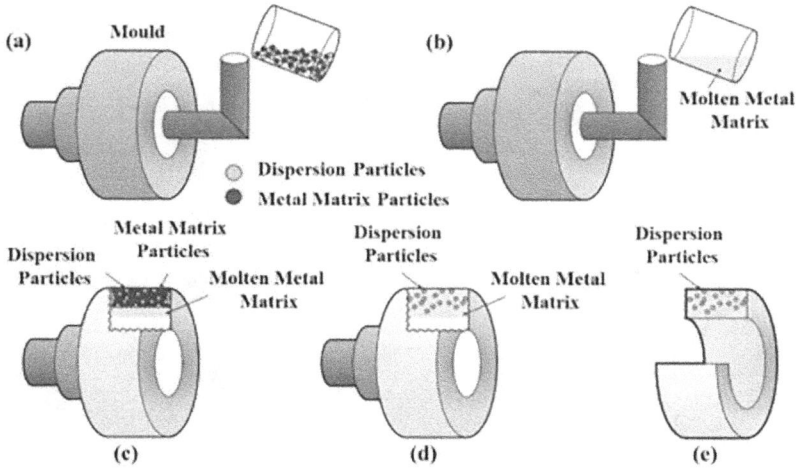

FIGURE 9.16 The centrifugal mixed power method produces FGM with continuous graded properties across thickness (Saleh, Jiang, Fathi et al. 2020). {permission has been taken}

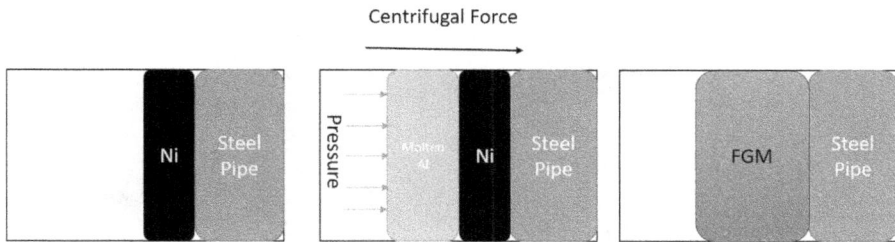

FIGURE 9.17 FG nickel-aluminides/steel multilayer pipe by reactive centrifugal casting method (Saleh, Jiang, Fathi et al. 2020).

9.3.21 TAPE CASTING

Multilayer structures, such as those found in multilayer condensers and circuits, are commonly cast using tape casting technologies. Doctor blading is a common form of the tape casting method (Cheng et al. 2018). As shown in Figure 9.18, a slurry of ceramic powder is continually poured onto a moveable layer of support made of flat non-bonding material such as Teflon, which contains an organic solvent such as ethanol and a variety of additional additives, such as a polymer binder (Nishihora et al. 2018).

9.3.22 LANGMUIR-BLODGETT METHOD

Langmuir-Blodgett (LB) film technology allows for the precise deposition of homogenous film materials up to a single-molecule layer thickness. The LB approach works

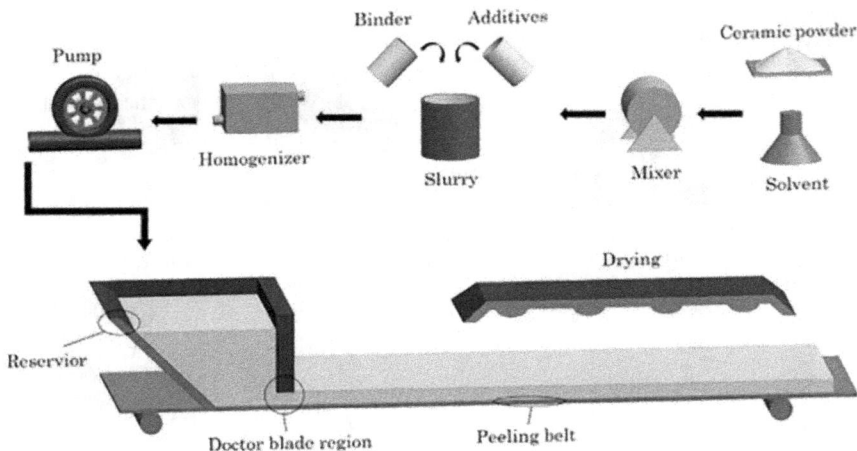

FIGURE 9.18   Tape casting process (Shanefield 1991). {permission has been taken}

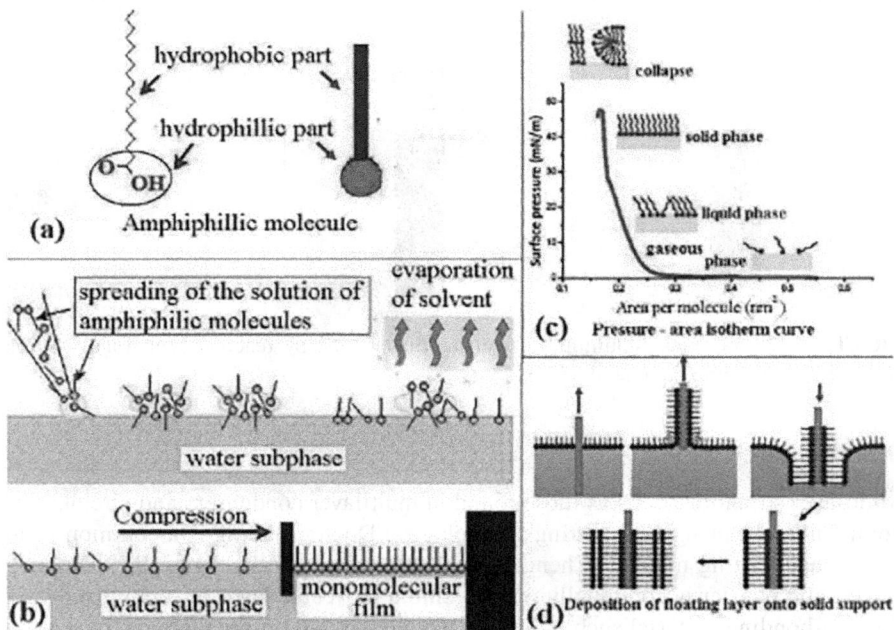

FIGURE 9.19   Schematic illustration of the Langmuir-Blodgett method (Hussain et al. 2018).

by forming an amphiphilic substance as a monomolecular coating on the water surface and then moving it to a solid substrate. The amphiphilic molecules are organized at the air-water contact during the aqueous stage. As seen in Figure 9.19, a unique membrane comprise of the surface layer to generate a single monomolecular surface layer.

The monomolecular film structure, transmitted via a collection of two-dimensional states known as water, is affected by consecutive isothermal compression.

As a result, having a film phase diagram allows to investigate the structure and physical-chemical characteristics of the films. The film is moved to a solid support, and a flat substrate is dipped in the solution and then drained with the adsorbed surface. A monomolecular film transfer method can be duplicated several times (Hussain et al. 2018).

9.4 SUSTAINABILITY ASPECTS OF FGM MANUFACTURING

Sustainable manufacturing (SM), or green manufacturing, can be described as a production approach that minimizes waste and environmental impact. These objectives will mostly be met by implementing techniques that will impact product design, process design, and operational principles. The energy required by conventional and AM processes for making products results in environmental damage (Bobba, Ardente, and Mathieux 2016; Ilgin and Gupta 2010; Z. Y. Liu et al. 2018; Previtali et al. 2020). The environmental impact is measured with Eco points; a score is given by considering various impact categories by considering CO_2 as the indicator for the global warming potential (GWP).

Sustainability analysis of the manufacturing process is performed by dividing the process according to the processing technique, as shown in Figure 9.20. Life cycle assessment (LCA) and specific energy consumption (SEC) are widely used for performing sustainability analysis of various manufacturing operations such as milling, shaping, and drilling (Depeursinge et al. 2010) presented a detailed comparison study between the AM technique and the conventional manufacturing process. The study was mainly focused on environmental issues. The analysis of all the processes, including production and preprocessing to the final product using LCA was discussed by the authors.

Manufacturing analysis will give an idea of the sustainability of the process using additive manufacturing or conventional processes. From Figure 9.21, it is clear that the energy consumption of the grinding process is the highest among all the subtractive processes. Milling also takes lesser SEC, as material is removed in the form of a bigger chip. The SEC of the turning and grinding processes is 15% and 20%, respectively. The figure shows the utilization of energy for the AM process. The FDM process constitutes 67% of the average SEC compared to SLS and SLA in the AM process. The SEC of the AM process is in KWh/Kg, which is significantly higher when compared to another conventional method. The AM process is a very energy-intensive process and consumes much energy and causes environmental pollution. The AM process is generally used in aerospace industries, for bio-implants, and in research where the product is not produced in mass. Compared to traditional machining, AM has a significant environmental effect, as seen in Figure 9.21. In addition, processing costs are the deciding factor in AM procedures since it involves many resources and produce a lot of GWP. The AM process is uneconomical based on the available data.

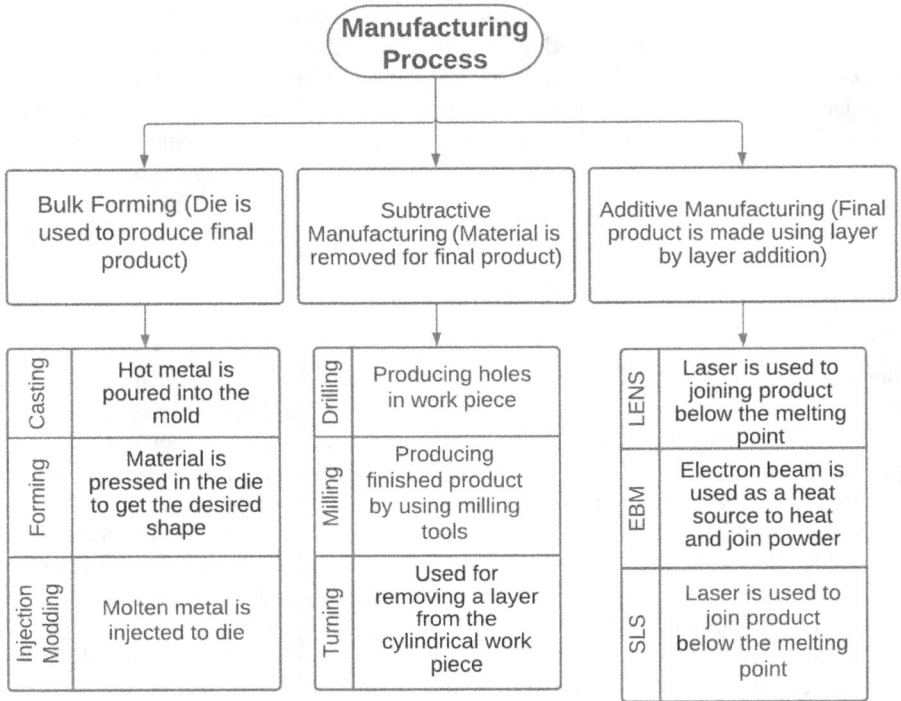

FIGURE 9.20 Division of the process based on different processing techniques.

FGM production methods can be broadly classified into low SEC, moderate SEC, and high SEC processes, as shown in Figure 9.22. Figure 9.22 shows a better choice of selecting manufacturing depending on the production requirement. The figure was developed using data from Z. Liu et al. (2016, 2018) and Previtali et al. (2020).

9.4.1 ECONOMICS OF PRODUCTION

In this section, economic aspects of production are discussed. Cheaper and less sophisticated process will be adopted for production. The cost and SEC of a product depend on the production quantity.

In injection molding, machining, and FDM (Yoon et al. 2014), it is clear that for the smaller parts, the injection molding process is a very energy-intensive process; the SEC decreases drastically with increased production., However, in the case of the FDM process, it has relatively higher SEC energy. However, it can be made economical if the quantity of the production less. This is a better option for smaller-scale production, but injection molding becomes economical for mass production.

Figure 9.23 uses data (El-Galy, Saleh, and Ahmed 2019a) showing that energy requirements differ for smaller and larger parts. The centrifugal casting process is the most efficient for small and large parts. The conventional production (method)

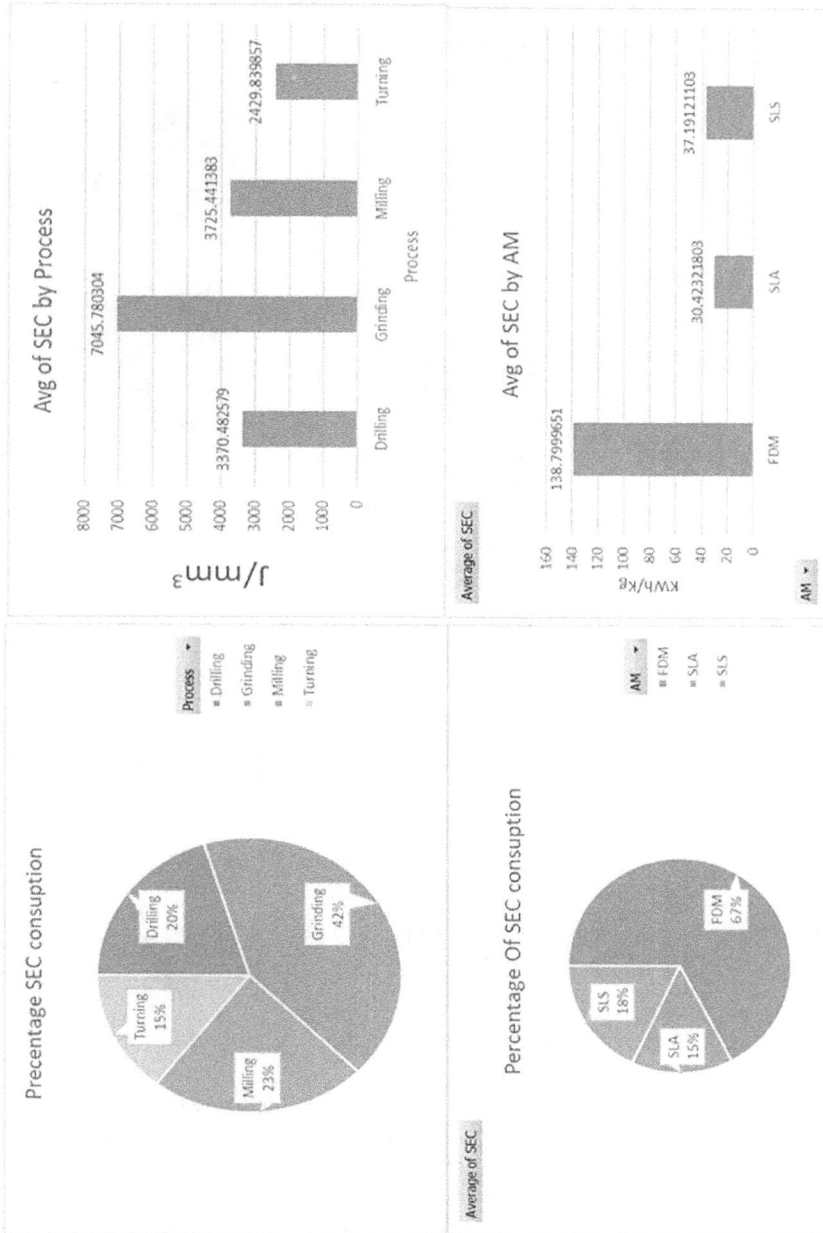

FIGURE 9.21 SEC Energy consumption of AM and conventional process (Z. Liu et al. 2016, 2018; Previtali et al. 2020).

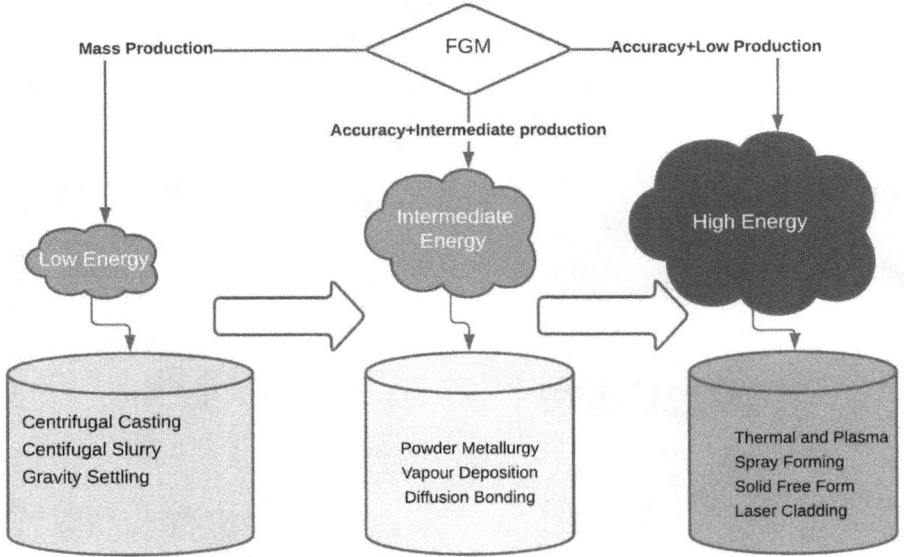

FIGURE 9.22   FGM manufacturing according to SEC.

FIGURE 9.23   Energy consumption of different processes with SEC.

mainly consists of centrifugal casting, powder metallurgy, and AM processes: SLS, SLM, and 3D printing.

Powder metallurgy and AM are most suitable for manufacturing FGM. Figure 9.24 shows the general understanding of the AM and conventional process application and the approach's feasibility. The figure indicates that the AM process is a very efficient solution for production when the output is lower and complexity is high. Conventional injection molding, PM, is mainly used for mass production and less complex product.

The economic impact and sustainability of the process determine its utility in the industry. The quantity of production, complexity of production, and type of material

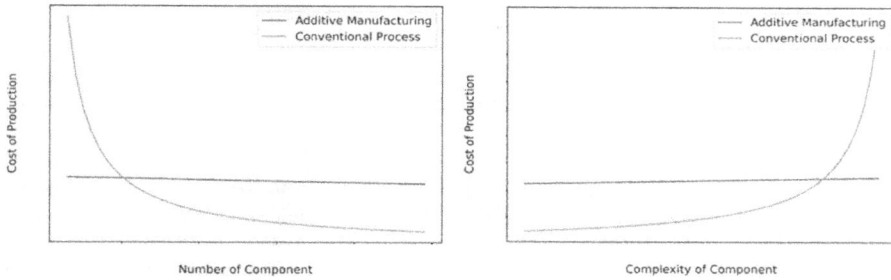

FIGURE 9.24 (a) Cost vs number of components; (b) cost vs number of components for additive and conventional manufacturing.

used should be considered for manufacturing. The optimum manufacturing process should be economical and eco-friendly with fewer Eco points.

CONCLUSION

FGM is already a proven advanced composite with various applications, such as reusable rocket engines, aerospace, and automobiles. In the present chapter, various manufacturing technologies, mathematical formulations, and applications have been reviewed comprehensively. The sustainability aspect of FGM has been discussed in detail. Despite its wide application, there is still a challenge in the manufacturability of FGM, such as bulk production and proper gradation. Recently, 3D printing has shown the capability of achieving specific gradations in FGM properties. However, concerns such as mass production adaptability, process repeatability, dependability, and cost-effectiveness are among the potential frontiers for FGM that have yet to be resolved.

REFERENCES

Benveniste, Y. 1987. "A New Approach to the Application of Mori-Tanaka's Theory in Composite Materials." *Mechanics of Materials* 6(2): 147–157.
Bever, M. B., and P. E. Duwez. 1972. "Gradients in Composite Materials." *Materials Science and Engineering* 10: 1–8. https://linkinghub.elsevier.com/retrieve/pii/0025541672900596
Bobba, Silvia, Fulvio Ardente, and Fabrice Mathieux. 2016. "Environmental and Economic Assessment of Durability of Energy-Using Products: Method and Application to a Case-Study Vacuum Cleaner." *Journal of Cleaner Production* 137: 762–776. http://dx.doi.org/10.1016/j.jclepro.2016.07.093
Burlayenko, V. N. et al. 2017. "Modelling Functionally Graded Materials in Heat Transfer and Thermal Stress Analysis by Means of Graded Finite Elements." *Applied Mathematical Modelling* 45: 422–438.
Chao, Z. L. et al. 2019. "The Microstructure and Ballistic Performance of B 4 C/AA2024 Functionally Graded Composites with Wide Range B 4 C Volume Fraction." *Composites Part B: Engineering* 161: 627–638.
Cheng, Laifei et al. 2018. "Structure Design, Fabrication, Properties of Laminated Ceramics: A Review." *International Journal of Lightweight Materials and Manufacture* 1(3): 126–141. https://doi.org/10.1016/j.ijlmm.2018.08.002

Choy, K. L. 2003. "Chemical Vapour Deposition of Coatings." *Progress in Materials Science* 48(2): 57–170.

Das, Mitun, Vamsi Krishna Balla, T. S. Sampath Kumar, and Indranil Manna. 2013. "Fabrication of Biomedical Implants Using Laser Engineered Net Shaping (LENS™)." *Transactions of the Indian Ceramic Society* 72(3): 169–174.

Deng, Chun, Hyeongwon Kim, and Hyungson Ki. 2019. "Fabrication of Functionally-Graded Yttria-Stabilized Zirconia Coatings by 355 nm Picosecond Dual-Beam Pulsed Laser Deposition." *Composites Part B: Engineering* 160: 498–504. https://doi.org/10.1016/j.compositesb.2018.12.101

Depeursinge, Adrien et al. 2010. "Fusing Visual and Clinical Information for Lung Tissue Classification in HRCT Data." *Artificial Intelligence in Medicine* 229: ARTMED1118.

Egbo, Munonyedi Kelvin. 2021. "A Fundamental Review on Composite Materials and Some of Their Applications in Biomedical Engineering." *Journal of King Saud University— Engineering Sciences* 33(8): 557–568. https://doi.org/10.1016/j.jksues.2020.07.007

El-Galy, Islam M., Bassiouny I. Saleh, and Mahmoud H. Ahmed. 2019a. "Functionally Graded Materials Classifications and Development Trends from Industrial Point of View." *SN Applied Sciences* 1(11): 1–23. https://doi.org/10.1007/s42452-019-1413-4

———. 2019b. "Functionally Graded Materials Classifications and Development Trends from Industrial Point of View." *SN Applied Sciences* 1(11). https://doi.org/10.1007/s42452-019-1413-4

Gandra, João Pedro Machado da. 2010. *Preliminary Study on the Production of Functionally Graded Materials by Friction Stir Processing.* Corpus ID: 134394973. http://hdl.handle.net/10362/4889

Gasik, Michael, Akira Kawasaki, and Sei Ueda. 2006. "Design and Powder Metallurgy Processing of Functionally Graded Materials." *Materials Development and Processing— Bulk Amorphous Materials, Undercooling and Powder Metallurgy*: 258–264.

Ghatage, Pankaj S., Vishesh R. Kar, and P. Edwin Sudhagar. 2020. "On the Numerical Modelling and Analysis of Multi-Directional Functionally Graded Composite Structures: A Review." *Composite Structures* 236.

Gupta, Ankit, and Mohammad Talha. 2015a. "Recent Development in Modeling and Analysis of Functionally Graded Materials and Structures." *Progress in Aerospace Sciences* 79: 1–14. http://dx.doi.org/10.1016/j.paerosci.2015.07.001

El Harti, Khalid et al. 2019. "Finite Element Model of Vibration Control for an Exponential Functionally Graded Timoshenko Beam with Distributed Piezoelectric Sensor/Actuator." *Actuators* 8(1).

Hazlehurst, Kevin Brian, Chang Jiang Wang, and Mark Stanford. 2014. "An Investigation into the Flexural Characteristics of Functionally Graded Cobalt Chrome Femoral Stems Manufactured Using Selective Laser Melting." *Materials and Design* 60: 177–183. http://dx.doi.org/10.1016/j.matdes.2014.03.068

Heimann, Robert B. 2007. Plasma-Spray Coating: Principles and Applications. *Plasma-Spray Coating: Principles and Applications.* Wiley Publishers. https://www.wiley.com/en-us/Plasma+Spray+Coating%3A+Principles+and+Applications-p-9783527614844

Hussain, Syed Arshad, Bapi Dey, D. Bhattacharjee, and N. Mehta. 2018. "Unique Supramolecular Assembly through Langmuir—Blodgett (LB) Technique." *Heliyon* 4(12): e01038. https://doi.org/10.1016/j.heliyon.2018.e01038

Ilgin, Mehmet Ali, and Surendra M. Gupta. 2010. "Environmentally Conscious Manufacturing and Product Recovery (ECMPRO): A Review of the State of the Art." *Journal of Environmental Management* 91(3): 563–591. http://dx.doi.org/10.1016/j.jenvman.2009.09.037

Jha, D. K., Tarun Kant, and R. K. Singh. 2013. "A Critical Review of Recent Research on Functionally Graded Plates." *Composite Structures* 96: 833–849.

Jiao, Lishi et al. 2018. "Femtosecond Laser Produced Hydrophobic Hierarchical Structures on Additive Manufacturing Parts." *Nanomaterials* 8(8).

Jung, Woo Young, and Sung Cheon Han. 2013. "Analysis of Sigmoid Functionally Graded Material (S-FGM) Nanoscale Plates Using the Nonlocal Elasticity Theory." *Mathematical Problems in Engineering* 2013. Hindawi Publicshers, https://doi.org/10.1155/2013/476131

Karandikar, Prashant et al. 2013. "Al/Al2O3 Metal Matrix Composites (MMCs) and Macrocomposites for Armor Applications." *37th International Conference and Exposition on Advanced Ceramics and Composites (ICACC)* ARL-RP-460 (September).

Kaya, C., F. Kaya, S. Atiq, and A. R. Boccaccini. 2003. "Electrophoretic Deposition of Ceramic Coatings on Ceramic Composite Substrates." *British Ceramic Transactions* 102(3): 99–102.

Khorsand, Mohammad, Kunkun Fu, and Youhong Tang. 2019. "Multi-Directional Functionally Graded Materials for Enhancing the Durability of Shell Structures." *International Journal of Pressure Vessels and Piping* 175(February): 103926.

Koizumi, M. 1997. "FGM Activities in Japan." *Composites Part B: Engineering* 28(1–2): 1–4. https://linkinghub.elsevier.com/retrieve/pii/S1359836896000169

Kumar Bohidar, Shailendra, Ritesh Sharma, and Ranjan Mishra. 2014. "Functionally Graded Materials: A Critical Review." *International Journal of Research (IJR)* 1(7).

Kumar, Subodh, and Prasun Jana. 2019. "Application of Dynamic Stiffness Method for Accurate Free Vibration Analysis of Sigmoid and Exponential Functionally Graded Rectangular Plates." *International Journal of Mechanical Sciences* 163(August): 105105. https://doi.org/10.1016/j.ijmecsci.2019.105105

Łatka, Leszek et al. 2020. "Review of Functionally Graded Thermal Sprayed Coatings." *Applied Sciences (Switzerland)* 10(15).

Liu, Z. Y., C. Li, X. Y. Fang, and Y. B. Guo. 2018. "Energy Consumption in Additive Manufacturing of Metal Parts." *Procedia Manufacturing* 26: 834–845. https://doi.org/10.1016/j.promfg.2018.07.104

Liu, Zhichao et al. 2016. "Energy Consumption and Saving Analysis for Laser Engineered Net Shaping of Metal Powders." *Energies* 9(10): 1–12.

———. 2018. "Investigation of Energy Requirements and Environmental Performance for Additive Manufacturing Processes." *Sustainability (Switzerland)* 10(10).

Maskery, I. et al. 2016. "A Mechanical Property Evaluation of Graded Density Al-Si10-Mg Lattice Structures Manufactured by Selective Laser Melting." *Materials Science and Engineering A* 670: 264–274. http://dx.doi.org/10.1016/j.msea.2016.06.013

Mathews, Nisha Grace, and Narasimha Murthy. 2017. *Analysis of Functionally Graded Material Plates Using Sigmoidal Law Energy Minimization in Cellular Networks View Project Security in WSN View Project.* www.researchgate.net/publication/319643868

Mori, T., and K. Tanaka. 1973. "Average Stress in Matrix and Average Elastic Energy of Materials with Misfitting Inclusions." *Acta Metallurgica* 21(5): 571–574.

Nejad, Mohammad Zamani, Negar Alamzadeh, and Amin Hadi. 2018. "Thermoelastoplastic Analysis of FGM Rotating Thick Cylindrical Pressure Vessels in Linear Elastic-Fully Plastic Condition." *Composites Part B: Engineering* 154: 410–422. https://doi.org/10.1016/j.compositesb.2018.09.022

Niknam, H., A. H. Akbarzadeh, D. Rodrigue, and D. Therriault. 2018. "Architected Multi-Directional Functionally Graded Cellular Plates." *Materials and Design* 148(2017): 188–202.

Nishihora, Rafael Kenji, Priscila Lemes Rachadel, Mara Gabriela Novy Quadri, and Dachamir
    Hotza. 2018. "Manufacturing Porous Ceramic Materials by Tape Casting—A Review."
    Journal of the European Ceramic Society 38(4): 988–1001. http://dx.doi.org/10.1016/j.
    jeurceramsoc.2017.11.047
Pan, Zengxi et al. 2018. "Arc Welding Processes for Additive Manufacturing: A Review." (2):
    3–24.
Pascon, J. P., and H. B. Coda. 2013. "High-Order Tetrahedral Finite Elements Applied
    to Large Deformation Analysis of Functionally Graded Rubber-like Materials."
    Applied Mathematical Modelling 37(20–21): 8757–8775. http://dx.doi.org/10.1016/j.
    apm.2013.03.062
Popoola, Patricia A.I., Gabriel Farotade, Olawale S. Fatoba, and Olawale Popoola. 2016.
    "Laser Engineering Net Shaping Method in the Area of Development of Functionally
    Graded Materials (FGMs) for Aero Engine Applications—A Review." *Fiber Laser*
    (April).
Previtali, B. et al. 2020. "Comparative Costs of Additive Manufacturing vs. Machining: The
    Case Study of the Production of Forming Dies for Tube Bending." *Solid Freeform
    Fabrication 2017: Proceedings of the 28th Annual International Solid Freeform
    Fabrication Symposium—An Additive Manufacturing Conference, SFF 2017*:
    2816–2834.
Rajak, Dipen Kumar et al. 2020. "Critical Overview of Coatings Technology for Metal
    Matrix Composites." *Journal of Bio- and Tribo-Corrosion* 6(1). https://doi.org/10.1007/
    s40735-019-0305-x
Reinhold, Ekkehart, J. Richter, U. Seyfert, and C. Steuer. 2004. "Metal Strip Coating by
    Electron Beam PVD—Industrial Requirements and Customized Solutions." *Surface
    and Coatings Technology* 188–189(1–3 SPEC.ISS.): 708–713.
Reisgen, U., Sharma, R., and Oster, L., 2019. "Plasma Multiwire Technology with
    Alternating Wire Feed for Tailor-Made Material Properties in Wire And Arc Additive
    Manufacturing." *Metals* 9(7): 745. https://doi.org/10.3390/met9070745
Ribeiro, Marcelo L. 2018. "A Delamination Propagation Model for Fiber Reinforced."
    Mathematical Problems in Engineering, Hindawi. https://doi.org/10.1155/2018/1861268
Rider, Patrick et al. 2018. "Additive Manufacturing for Guided Bone Regeneration: A
    Perspective for Alveolar Ridge Augmentation". *International Journal of Molecular
    Science* 9(11): 3308. https://doi.org/10.3390/ijms19113308.
Rodrigues, Tiago A. et al. 2019. "Current Status and Perspectives on Wire and Arc Additive
    Manufacturing (WAAM)." *Materials* 12(7).
Saleh, Bassiouny, Jinghua Jiang, Reham Fathi, et al. 2020. "30 Years of Functionally
    Graded Materials: An Overview of Manufacturing Methods, Applications and Future
    Challenges." *Composites Part B: Engineering* 201(August): 108376. https://doi.
    org/10.1016/j.compositesb.2020.108376
Saleh, Bassiouny, Jinghua Jiang, Aibin Ma, et al. 2020. "Review on the Influence of
    Different Reinforcements on the Microstructure and Wear Behavior of Functionally
    Graded Aluminum Matrix Composites by Centrifugal Casting." *Metals and Materials
    International* 26(7): 933–960. https://doi.org/10.1007/s12540-019-00491-0
Saxena, Abhinav, Shivam Gupta, Bhupendra Singh, and Ashutosh Kumar Dubey. 2019.
    "Improved Functional Response of Spark Plasma Sintered Hydroxyapatite Based
    Functionally Graded Materials: An Impedance Spectroscopy Perspective." *Ceramics
    International* 45(6): 6673–6683. https://doi.org/10.1016/j.ceramint.2018.12.156
Shanefield, D.J. 1991. "Tape Casting." *Concise Encyclopedia of Advanced Ceramic Materials*:
    469–472.

Sharma N.K, Bhandari M, and Ashirvad. 2014. "Applications of Functionally Graded Materials (FGMs)." *International Journal of Engineering Research & Technology (IJERT)*: 334–339.

Singh, Dheer, and Ankit Gupta. 2022. "Materials Today : Proceedings Influence of Geometric Imperfections on the Free Vibrational Response of the Functionally Graded Material Sandwich Plates with Circular Cut-Outs." *Materials Today: Proceedings* (xxxx): 2–5. https://doi.org/10.1016/j.matpr.2022.02.187

Singh, Rudra Pratap, Somil Dubey, Aman Singh, and Subodh Kumar. 2020. "A Review Paper on Friction Stir Welding Process." *Materials Today: Proceedings* 38(June): 6–11.

Sobczak, Jerzy J., and Ludmil Drenchev. 2013. "Metallic Functionally Graded Materials: A Specific Class of Advanced Composites." *Journal of Materials Science and Technology* 29(4): 297–316. http://dx.doi.org/10.1016/j.jmst.2013.02.006

Tanaka, K. et al. 1993. "An Improved Solution to Thermoelastic Material Design in Functionally Gradient Materials: Scheme to Reduce Thermal Stresses." *Computer Methods in Applied Mechanics and Engineering* 109(3–4): 377–389.

Thai, Huu Tai, and Seung Eock Kim. 2015. *128 Composite Structures: A Review of Theories for the Modeling and Analysis of Functionally Graded Plates and Shells.* Elsevier Ltd.

Toudehdehghan, Abdolreza et al. 2017. "A Brief Review of Functionally Graded Materials." *MATEC Web of Conferences* 131: 1–6.

Wang, Jun et al. 2018. "Characterization of Wire Arc Additively Manufactured Titanium Aluminide Functionally Graded Material: Microstructure, Mechanical Properties and Oxidation Behaviour." *Materials Science and Engineering A* 734: 110–119. https://doi.org/10.1016/j.msea.2018.07.097

Watanabe, Yoshimi, Hiroyuki Eryu, and Kiyotaka Matsuura. 2001. "Evaluation of Three-Dimensional Orientation of Al_3Ti Platelet in Al-Based Functionally Graded Materials Fabricated by a Centrifugal Casting Technique." *Acta Materialia* 49(5): 775–783.

Yan, Jingyuan, Ilenia Battiato, and Georges M. Fadel. 2018. "Planning the Process Parameters for the Direct Metal Deposition of Functionally Graded Parts Based on Mathematical Models." *Journal of Manufacturing Processes* 31: 56–71. http://dx.doi.org/10.1016/j.jmapro.2017.11.001

Yoon, Hae Sung et al. 2014. "A Comparison of Energy Consumption in Bulk Forming, Subtractive, and Additive Processes: Review and Case Study." *International Journal of Precision Engineering and Manufacturing—Green Technology* 1(3): 261–279.

Zenkour, Ashraf M. 2006. "Generalized Shear Deformation Theory for Bending Analysis of Functionally Graded Plates." *Applied Mathematical Modelling* 30(1): 67–84.

Zhang, Chi et al. 2019. "Additive Manufacturing of Functionally Graded Materials: A Review." *Materials Science and Engineering A* 764: 138209. https://doi.org/10.1016/j.msea.2019.138209

10 Performance Enhancements Using Lean Manufacturing Practices in SMEs

A Case Study from Northern India

Sachin Saini, Doordarshi Singh, Ishbir Singh

CONTENTS

10.1 INTRODUCTION

Lean manufacturing is attractive to manufacturing industries in the 21st century due to its numerous benefits, such as low-cost products, shorter cycle times, timely deliveries, fewer rejections and fewer customer complaints (Bai et al., 2019; Chauhan and

DOI: 10.1201/9781003269298-10

Chauhan, 2019; Saini and Singh, 2021). Lean manufacturing identifies and eliminates non-value-added activities throughout the supply chain (Rameez and Inamdar,
2010; Saini and Singh, 2022b). Deriving its name from the Toyota production system (TPS), manufacturers started to practice elements of the lean manufacturing
approach such as just in time (JIT), Kanban, setup time reduction, production level-
ing and quality circles (Mohanty et al., 2006; Saini and Singh, 2022a). And operational benefits obtained by implementing lean manufacturing approach were lower
inventory levels, utilizing maximum space and minimum changeover time (Saini
and Singh, 2022c; Naveen et al., 2013). Its implementation in small and medium
enterprises (SMEs) is always a major concern for manufacturers (Yadav et al., 2019)
because SMEs operate with minimal resources and depend upon few customers.
Lean manufacturing is an approach in which organizations can make the most of
existing resources (Belhadi et al., 2019). Lean manufacturing practices (LMPs) are
also beneficial, as these can contribute to the economy of the nation for larger interests. This chapter is structured as follows: the next section sheds light on the literature of successful performance of lean manufacturing in various contexts. In Section
10.3, the research problem and introduction to the case study organization are given.
Section 10.4 portrays the enhancements by adopting LMPs in the case study organization. Section 10.5 deals with the discussion and conclusions regarding the benefits
obtained by using LMPs on the shop floor.

10.2 LITERATURE REVIEW

**Contribution of lean manufacturing practices to enhancing business perfor-
mance:** Lean manufacturing is a set of practices that target eradicating waste from
the system and improving the flow in the system. Value stream mapping (VSM)
is a lean practice used as an improvement technique to enhance the material flow
by the visualization of the entire production process (Sullivan and Aken, 2002). In
another study, industries embracing LM systems and alterations in plant layout led
to a shorter distance traveled for material to be dispatched (Gurumurthy and Kodali,
2010). Reducing setup time and implementing JIT increased productivity from 222
to 400 per operator (Ramnath et al., 2010). The production of the pump set rose
from 3200 to 8000 in another study by adopting lean tools (Rameez and Inamdar,
2010). Eradicating waste from the system, a hike in productivity and cost savings of
around 9.6 lakhs per annum were reported in another study by adopting lean tools
(Rajenthirakumar and Thyla, 2011). Product development time of around 33% was
saved with the lean manufacturing approach (Parkash and Kumar, 2011). Setup time
dropped from 40 to 12 minutes, and the production rate increased with the implementation of single minute exchange of dies techniques (Kumar and Abuthakeer,
2012). Machine down time was reduced to 14.17% compared to the original processing method (Ruhani and Muhammad, 2012). The Kaizen tool improved flow in
manufacturing lines, and as a result, productivity improved there (Mehta et al., 2012).
After going through the literature review, it is found that LMPs are a proven methodology to enhance performance parameters (Saini and Singh, 2022c). Apart from
achieving operational benefits by using LMPs in SMEs and its effects on sustainable
developments also has been depicted in this study.

10.2.1 LINKING LEAN MANUFACTURING PRACTICES WITH SUSTAINABLE DEVELOPMENTS

Thomas et al. (2012, p 428) defines sustainable development as "development that meets the needs of the present without compromising the ability of future generations to meet their own needs." It is related to the economic, social, institutional and environmental aspects of human society. In the field of the lean manufacturing, a number of studies were found citing the credibility of sustainable performance using LMPs. Manufacturing practices were designed with the motive of reducing waste, optimizing use of resources and improving workforce competencies to enhance process efficiency (Saini and Singh, 2018a). LMPs is a concept developed by Toyota at the end of the 20th century for manufacturing products with zero waste. Its primary target is to banish waste anywhere in the system of the organization. With the removal of waste from the system, a number of things are conserved; resources are preserved, there are fewer emissions, a pollutant-free environment is created and it makes the planet healthier, all apart from the operations benefits (Piercy and Rich, 2015). Sustainable development is also designed to focus on three aspects: socio (people), economic (profit) and environmental (planet), for the betterment of humankind. Bergenwall et al. (2012) studied the effect of the Toyota process design in American automotive plants on the triple bottom line and its effect on sustainability. Pakdil and Leonard (2017) also advocated for lean implementation and focusing on sustainability to enhance firm success in attaining priorities. In a study of Malaysian automotive makers, LMPs were found to have a mediating effect between ISO 14001 and environmental performance (Habidin et al., 2018). This study also pointed out the common enemy, waste, in lean and green management. Another study emphasized adopting lean culture in a manufacturing firm in Malaysia to enhance sustainable performance (Iranmanesh et al., 2019). A study by Caldera et al. (2019) in Australia pointed to combining lean thinking with green practices for improving environmental performance as sustainable business practices. In the Indian context, in studies by Deshmukh et al. (2010) and Sajan et al. (2017), it was given thrust to go for LMPs for achieving sustainable developments and hence sustainability especially focusing on SMEs.

10.3 DRIVING FORCES AND THRUST BEHIND THIS STUDY

Today, the manufacturing industry has to respond rapidly to new demands and compete in a continuously changing environment, seeking out new methods to allow it to remain competitive and flexible simultaneously (Singh et al., 2011). The aim of manufacturing organizations is to reduce manufacturing costs and was through system simplification, organizational potential and proper infrastructural planning by using modern techniques like LMPs (Saini and Singh, 2020a).

In India, large numbers of medium- and large-scale manufacturing companies are dependent upon SMEs for manufacturing of vital parts used in assembly lines. The operations of SMEs differ from larger enterprises in number of ways (Upadhye et al., 2010). Keeping in view these facts, different tools of LMPs can benefit firms in the near future. So, the present work is focused on the material parts handling industry to see the actions taken to improve its overall performance. There are a number of lean manufacturing tools available, but experience and complete knowledge of manufacturing processes are required to select the appropriate tool for a specific process (Saini and Singh, 2020b). The standard selection and implementation procedure of lean tools must be adopted for successful removal of waste with minimum cost and time (Saini and Singh, 2022a). In the present chapter, selected lean practices were studied to investigate performance enhancements through lean practices in the case study organization (Saini and Singh, 2018a).

Objectives:
1. To review the current situation of the case study by SAP (Situation, Action and Process) analysis and LAP (Learning, Actor and Performance) implementation.
2. To enumerate performance enhancements by LMPs.

10.3.1 CASE STUDY

The present manufacturing organization is called VED Ltd. (anonymized) and is an SME located in northern India. It is a manufacturer of material handling parts for utility vehicles and trucks. The organization has been in operation for more than 15 years and has 55 employees. Globalization has caused some recent changes in the business originating from fluctuations in the market and the loyalty of the customers toward cost-effective products. All these factors added stress for the manufacturers, who are working with conventional methodologies. This can cause customers to shift their loyalty to other companies for better products at a better price. All these circumstances forced the manufacturer to go for lean practices to retain their loyal customers and enhance their firm performance. This organization has an annual turnover of 30 crores, with 55 employees, and it is spread over 15 acres. The present study focuses on the process of implementing LMPs on their workshop floor.

10.4 METHODOLOGY

In this case study, the manufacturing organization VED Ltd. decided to review the existing process and follow the implementation steps to improve the systems in the organization. In the next step, a methodology was developed for analyzing the existing system and developing future plans with the help of lean experts. A detailed step-by-step plan was prepared by experts in discussion with the top management. In this case, SAP and LAP analysis was conducted before and after implementing lean practices in the organization, depicted in Table 10.1. SAP is analysis done with motive of reviewing the existing procedures, and LAP extracts learning for further enhancements (Upadhye et al., 2010). After that, the action is implemented, and an audit is conducted at the end of three months for follow-up on lean initiatives (Vienazindiene and Ciarniene, 2013). Some of the improvement projects done by implementing LMPs in the organization are discussed in the following.

TABLE 10.1
SAP Analysis of the Organization

Situation	Action	Process
Customer pressure for JIT deliveries	Managing director	Improving processes for competitive spirit
Qualitative products	Top management	Implementing lean manufacturing practices
Competitive price	Skilled personnel	Use of statistical tools for reducing variance
Customer base	Quality	Training program enhancement
Tier-1 supplier to major automotive customers	Image in market	Improved work environment
		Communication and computerization enhanced
		Literacy for computerization increased
		Total employee involvement
		Creating belongingness among employees for workplace

After the SAP analysis, various issues emerged from the workshop floor:

- Wastage of time in tracing things
- Defects
- Tool costing
- Inventory
- Overall Equipment Effectiveness (OEE) enhancement
- Customer complaints

Looking at various issues that emerged from the process, practices and workshop floor, it was necessary to target those practices that would eliminate defects and enhance the overall performance of the organization. So the organization decided to embrace LMPs, which are discussed further in brief.

10.4.1 IMPLEMENTATION OF LEAN MANUFACTURING PRACTICES IN THE CASE STUDY ORGANIZATION

10.4.1.1 5S Practices

LMPs encompass a variety of tools, but their selection remains a big worry, especially for small manufacturers. SMEs operate with minimal resources and are dependent on a few customers. Then the initiative of enhancements starts from where, and it pinpoints towards adoption of the basic lean practice of 5S (i.e., seiri, seiton, seiso, standardize and shitsuke) (Antony, 2011; Rose et al., 2011; Bai et al., 2019). It also lays the foundation for any other practices to be established in the organization, and its universal applicability attracts manufacturers even for pilot projects (Abdulmalek and Rajgopal, 2006). Its implementation in the system eradicates hidden waste, which in turn is a progressive step toward sustainable development in the organization. Its enhancement on the workshop floor leads to reduction in non-value-added activities and streamlines the workplace. This practice is divided into five subsections: seiri, seiton, seiso, standardize and shitsuke. Before implementation of LMPs, a 5S audit was done to assess the current level of practice on the shop floor, and the score was 32. This was a bit low and not a good performance on the shop floor. Some of the reasons for the low 5S score were the absence of regular audits and daily routines for adopting this at the core level. Then, for effectiveness on the workshop floor, the standard operating procedure (SOP) manual was translated into the local language, and training programs on this practice were enhanced (Figure 10.4; Parry and Turner, 2006). After six months, another audit of the 5S score was done to assess the performance level, and the score was 58 (Table 10.3). Although a score of 58 is not in the satisfactory range for improvement purposes, it is indicative of management efforts from learning perspective (Figure 10.7; Gupta and Jain, 2015; Singh and Ahuja, 2015). It can be clearly seen that 5S implementation enhances the traceability and availability of the right things in the right place, so the delivery rate improves in the organization.

For the proper adoption of LMPs in the production area, management held a meeting with the supervisor, quality head, production manager and experts in the field. From this session, it was decided to implement 5S in full-fledged documented format. Some of the other benefits obtained, like coding of almirahs and streamlining the workshop area, are depicted in Figures 10.2 and 10.3. From Figure 10.5, training programs for workers were enhanced from a safety point of view. Also the implementation of 5S also impacted the delivery line, where finished goods were placed in a systematic manner (Figure 10.6). Some of the other reasons for SMEs to initiate lean manufacturing from the 5Ss are lower cost, less time-consuming processed and ease of implementation. Some other authors also argued for the application of 5S as a starting point for improving productivity and attaining high goals and priorities (Singh and Ahuja, 2015; Gupta and Jain, 2015; Veres et al., 2018). Some noteworthy improvements recorded after implementing 5S are mentioned in the following.

Time saving in tracing things on shop floor after 5S implementation:

Before 5S, time wasted in tracing things = 22 minutes (in one shift of 8 hours)
After 5S, time taken for tracing things = 7 min
Labor cost = 9500 per month, = 9500/30 × 8 × 60 = 0.659 Rs/min (for 8-hour shift)
Time saved = 22 − 7 = 15 minutes (one shift) = 30 min (two shifts)
Cost analysis = 30 × 1.32 = 39.58 Rs (two shifts) = 39.58 × 30 (for 1 month) = 1187.49/Rs per month or 1187.49 × 12 = 14,249.99 Rs (1 year)

In terms of cost analysis, the organization saved 1.5 times the labor cost per month. After this, the organization also introduced red flags in the production and administrative departments. This helped in visualizing the flow of material in the system, and red flags indicated critical areas. In the production system, each product goes under a number of transformations to reach the finishing stage, and this color coding helps management trace the material and recognize the weaker sections, depicted in Table 10.2.

TABLE 10.2

Identification of Red Flags in the Daily Progress Report

Identification of Red Flags in the Daily Progress Report						
CNC	VMC	Quality	Machining	E-27/Profile	Incoming	Admin
	Smooth					
	Critical					
	In time					

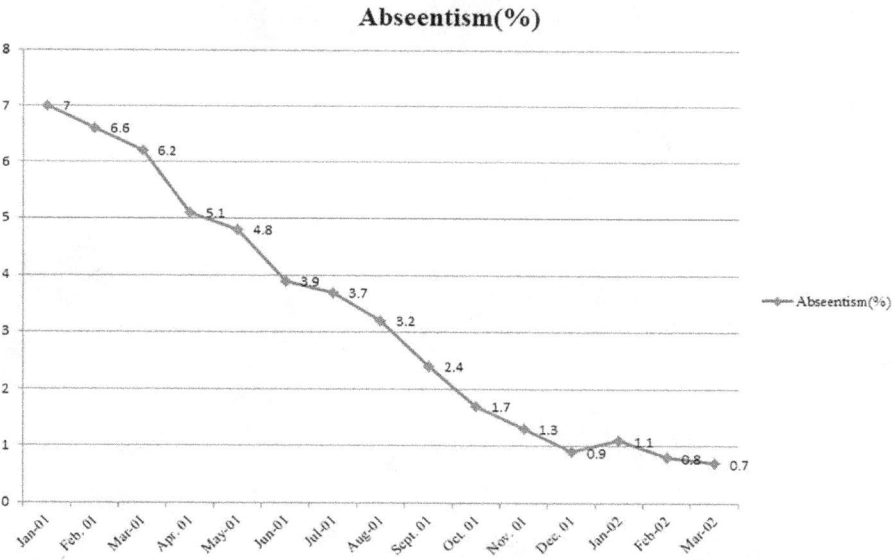

FIGURE 10.1 Absenteeism record.

The aftermath effects of implementing 5S on worker absenteeism were also recorded, and the trend indicated that absence from the workplace decreased, depicted in Figure 10.1. This may be because the awareness spread by the training program led to enhanced employee faith in the organization system.

10.4.2 SUPPLIER MANAGEMENT (PERFORMANCE SCORE)

In this practice, the organization assesses the current performance of the existing suppliers and evaluates them in terms of raw material supplied, in-process rejections

FIGURE 10.2   Coding of racks.

FIGURE 10 3   Workshop area.

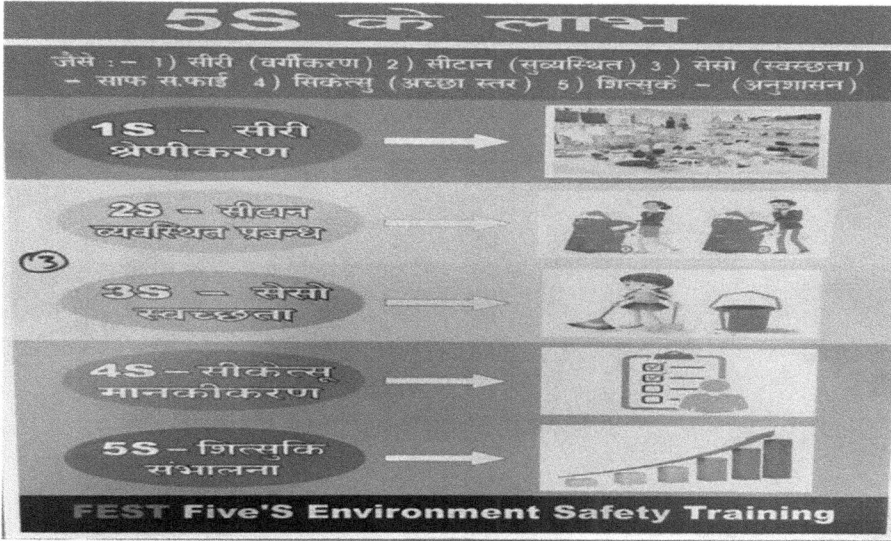

FIGURE 10.4 5S SOP.

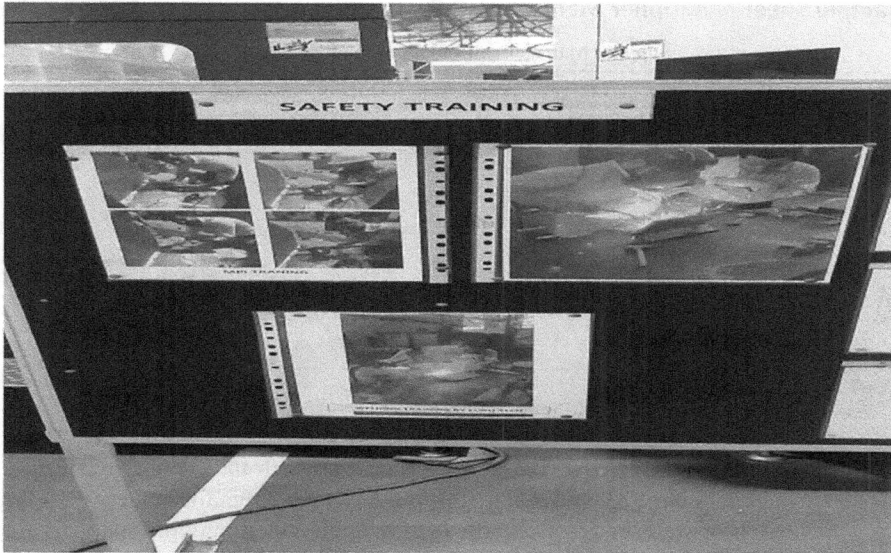

FIGURE 10.5 Training programs.

FIGURE 10.6 Finished goods.

TABLE 10.3
Sample Sheet of Supplier Management Practices

Supplier Management Practices (Sample Sheet)

S No.	Supplier Code/ Name	PO Quantity	Target Date	Actual Date	Day	Quality Accepted	Rejected	Rating/ Performance Score	Grade
1	VXZ Metals	23000 (Mild Steel plates)				17495	5560	76%	B
2	FRV Steels	8000 Mild Steel bars (25 × 175)				8000	0	90%	A

and delivery time by suppliers. These are assessed through the supplier evaluation sheet depicted in Table 10.3, and bonding through suppliers remains intact. With the implementation of supplier practices, the organization added multiple suppliers for various components to deliver the customized products in time. These evaluations kept performance pressure on the suppliers to remain in a race of competitiveness to supply raw materials on time at competitive prices. These days, where the pace of production is directly correlated to the takt time which creates consistent stress on manufacturers mind to keep the optimized inventory also thrust on improving supplier practices. Before implementing this practice, the organization stored inventory for 120 days in advance; after this implementation, the inventory period was reduced to 15–30 days. These changes created the addition of new customers, where quicker supply of products was a valuable contribution to the image of the organization (Saini and Singh, 2018b).

10.4.3 STATISTICAL QUALITY CONTROL (RATE OF REJECTION)

Quality is a vital characteristic of a product under consideration for sale in the market. Today, demand is for cost-effective products but at qualitative prices. So, for manufacturing products, considerations may vary for manufacturers and consumers. All these points focus on improving the process and refining the quality of products. From the array of lean practices, statistical quality control encompasses bar graphs, Pareto diagrams, process capability index and Poka Yoka methods for improving the quality of the process (Upadhye et al., 2010). First-pass correct output was also the driver behind lean practices in another study (Ghosh, 2013). The organization started to implement why-why analysis on the seeing the rejections on the shop floor depicted in Figure 10.8. Why-why analysis targets five Ms: men, money, materials, methods and machines. After applying this procedure, the root of the problem was identified: less skilled operators (men). This reduced the number of rejections on the shop floor, and calculating these in ppm brought a considerable amount of savings. In Table 10.4, reject quantity is depicted against the total inspected quantity, and calculating its ppm presents improvements in quality. After the implementation of statistical practices in the case study organization, an attempt was made to identify

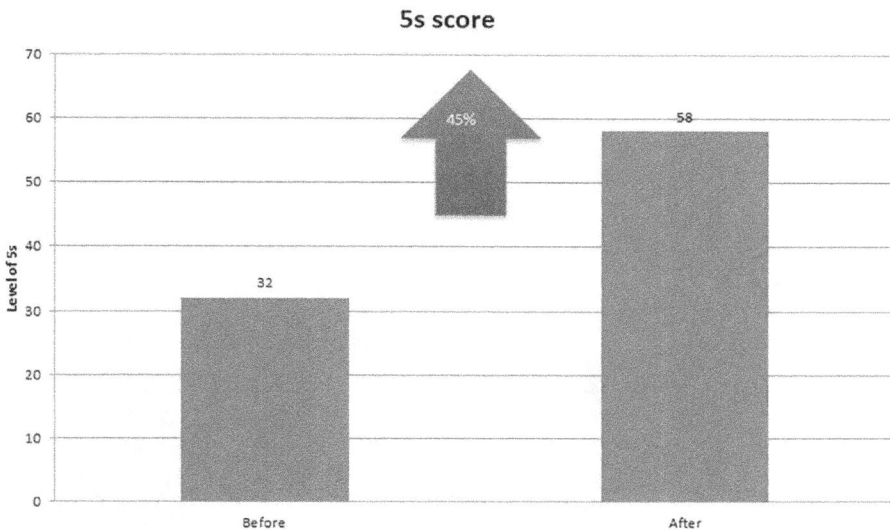

FIGURE 10.7   5S audits before and after implementation of LMPs.

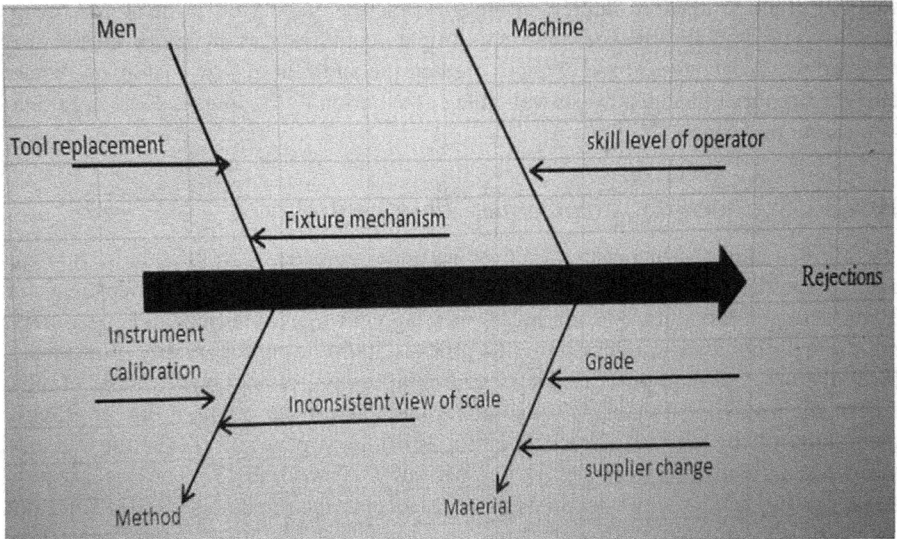

FIGURE 10.8   Fishbone diagram identifying the cause of rejections.

TABLE 10.4
Rejections by Month

Month	Total Inspect Quantity	Total Reject Quantity
Jan-01	3989	39
Feb-01	3307	44
Mar-01	4602	42
Apr-01	3753	41
May-01	3649	34
Jun-01	4465	22
July-01	3916	15
Aug-01	4550	9
Sept-01	4582	14
Oct.-01	1717	10
Nov.-01	2196	7
Dec-01	3020	8

the exact cause of rejections. The cause of rejections was found to be an unskilled operator performing the job, depicted in Figure 10.8. Management trained the operator to enhance process efficiency. So, the impact of lean on rejections was seen, and a reduction in rejections was achieved by the organization in the first year of implementing lean practices (Table 10.5).

10.4.4 Kaizen Practices

Kaizen are lean practices specifically aimed at continuous improvements. The word "kai" means continuous, and "zen" means improvements. In the absence of continuous improvement programs in the industries, lack of willpower of management for surviving and thriving in these times is seen. Kaizen practices are short-terms program designed to maximize the efficiency of the process and to improve the overall performance of the firms. In the case study organization, two continuous improvements programs were done, one saving in tool consumption and other one inventory management to maximize the efficiency of the organization, explained in the following.

10.4.4.1 Saving in Tool Consumption

Another problem found in the case study organization was the excessive tool consumption found in the machine shop. It was decided to change the tool bit, and a slight reduction in tool consumption started to appear. After brainstorming sessions by the management with the plant head, production supervisor and machine operator, a new tool bit was introduced for better results. So, as expected, the problem of excessive tool consumption was completely solved in 6 months (Table 10.5). In the same year, the organization decided to implement an optimized inventory level for consumables on the shop floor for various production purposes in the form of Excel sheets, presented in Table 10.6. Seeing the effectiveness of inventory management, the organization planned to purchase enterprise resource planning (ERP) systems in the near future.

TABLE 10.5
Consumption of Tool Bit by Month

Period	Consumption (in Rs)
Jun	350274
July	374940
Aug	338020
Sept.	298621
Oct.	201729
Nov.	176718
Dec.	157187

TABLE 10.6
Inventory Management for Various Tools Bits

S. Number	Material Description	Opening Stock (Year 01)	Purch	Total Quantity	Total Issue Quantity	Closing Stock	Min.	Max.
1	APMT 1604PDER-M2	40		40	8	32	30	60
2	BDMT 170408ER-JT	21	20	41	6	35	20	40
3	CNMA 120408 UC 5115	18	40	58	37	21	30	70
4	R 290–12T308M-PM	66		66	0	66	30	70
5	SNXX 1506 ANTN-M	1	10	11	0	11	5	20

10.4.5 PEOPLE PRACTICES (TRAINING PROGRAMS)

Training programs are vital for successfully infusing newer methodologies into the mindset of the people on the ground floor. These programs are designed for refining the process involved with upgrading the skills and methods of employees. Employee initiatives can pace the development works on the shop floor of the organization if trained upto that level (Sahoo et al., 2011; Knol et al., 2019). This will not be possible unless a vision or goal is set up by the management. SMEs run on orders from large enterprises but are resource constrained in terms of men, money and methods. Especially in case of the SMEs, these training programs are crucial for bringing about transformations in the organization. In this study, training programs were introduced fortnightly, and a monthly check sheet was implemented with the coordination of the management representative to effectively implement lean practices in the organization (Figure 10.9).

10.4.6 PREVENTIVE AND MAINTENANCE BREAKDOWNS

Cycle time and takt time both have a significant contribution to enhancing the performance of the process prevailing in the organization. For this, breakdowns and downtime period act as barriers to enhancing production the on shop floor (Upadhye et al., 2010). These things can be completely avoided by implementing a preventive maintenance policy and corrective actions on the shop floor. In this

FIGURE 10.9 Worker performing duties after safety training.

practice, a glance at various ongoing maintenance practices organization studied for gauging the level of maintenance work in the organization. The area of maintenance was pointed out for adapting to lean in Chinese firms (Taj, 2008). A policy of preventive maintenance was implemented in consultation with lean experts. After implementing this policy, the organization recorded considerable time savings to enhance production (Figure 10.10 and Table 10.7). Seeing this significant change, the top management decided to implement total predictive maintenance (TPM) in the next phase.

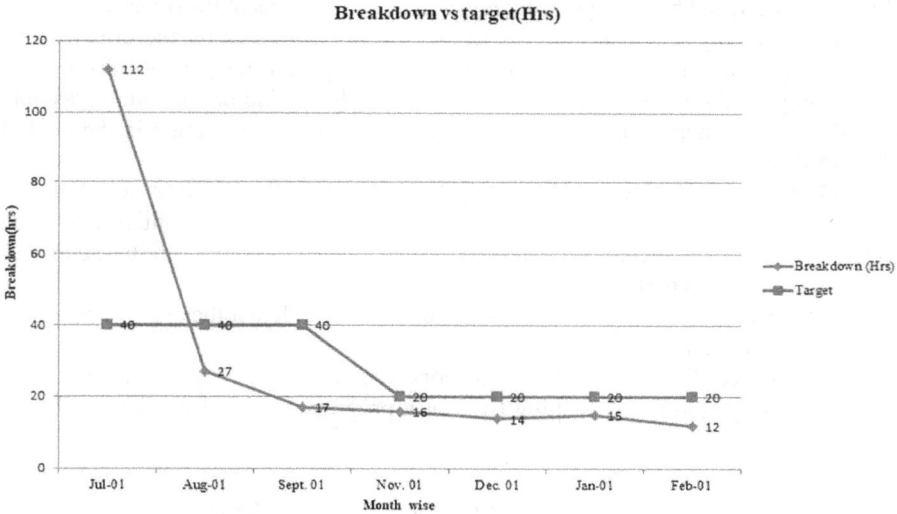

FIGURE 10.10   Depiction of breakdown vs. target (hrs) for maintenance schedule.

TABLE 10.7
Breakdown Hours vs. Target

Month	Breakdown (Hrs)	Target
Jan	112	40
Feb	27	40
March	17	40
April	16	20
May	14	20
Jun	15	20
July	12	20

10.4.7   ISO Certification (Plant Management Practices)

After implementing LMPs in a fragmented manner for the first year, the top management planned to adopt a quality management system (QMS) in the organization. Some of the motives for adopting a QMS were regular quality check, maintaining documentation, regular audits, tab on wastage and customer satisfaction (Jain and Samrat, 2015). So, the organization decided to implement ISO certification 9001:2015 after implementing LMPs in the next year. This was necessary to bring about transparency in system operations and enhance the customer attention towards ISO-certified products. This was being done after a meeting of top management, lean experts and ISO auditors. After this, the processes were going to record the form of documentation, and hence regular monitoring of the progress of the organization was achieved. Looking at enhanced work for documenting records for the organization and the organization decided to divide the work among qualified person on shop floor for keeping all the things in a streamline manner there. The organization was able to add new customers, apart from fluctuating orders after certification in the market (Figure 10.11).

LAP analysis: Learning: identifying the issues and their causes, top management commitment and support, waste identification, expansion capabilities, employee involvement and enhancing competencies, imparting lean in daily routines, communication from top to bottom.

Action: Steps taken by the management of VED Ltd. depict the ability and vision to transform into a lean organization.

Performance: Overall learning can conclude that designing a simple performance measure system for keeping track at a progress level will keep them in the hunt for survival.

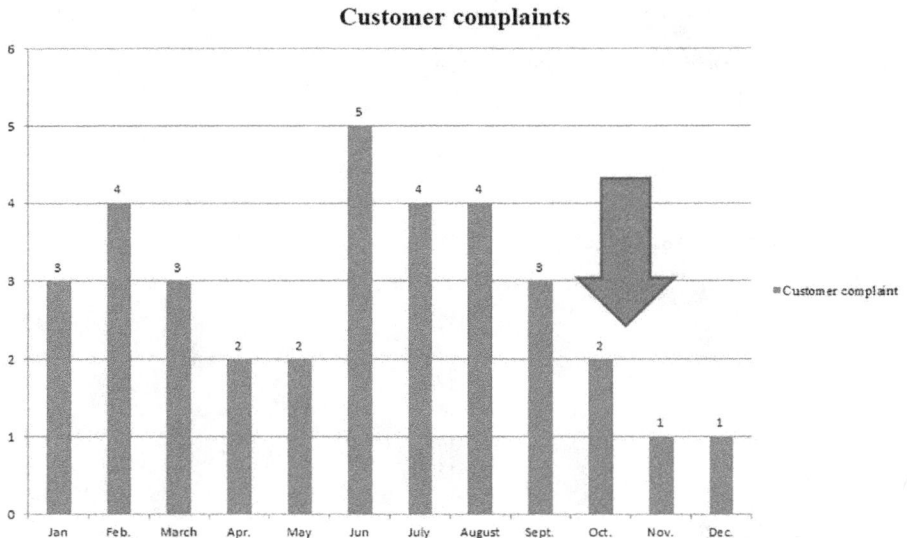

FIGURE 10.11   Reduction in customer complaints after introduction of LMPs.

Factors to be included in performance measurement system:

- Cost savings (reductions)
- Productivity
- Delivery rate
- OEE
- Rejections (drawing comparisons)
- Overall firm performance

10.5 DISCUSSION AND CONCLUSIONS

This case study showed a picture of implementing LMPs in an organization that was under stress due to customer pressure. The implementation of lean practices in SMEs is dependent upon a number of factors. LMPs require management vision and preplanning to implement. In this study, the implementation of LMPs started from the 5S audit score on the shop floor of the organization and laid the foundation for other lean practices. The overall firm performance improved after implementing lean practices, presented in Table 10.8, within a time span of 15–18 months (Yadav et al., 2019). Some of the benefits obtained by implementing lean practices were reduction in rejections, minimizing tooling consumption, inventory reduction and a decline in customer complaints. Training programs enhanced the competency of the people, and maintenance policy improved the efficiency of the process. Overall it can be concluded that SMEs focus on those lean practices that are easy to implement and have less investment involved. This is in line with other studies (Bai et al., 2019; Rose et al., 2011). Other conclusions drawn from the study are as follows:

- LMPs enhance firm performance if initiatives are taken by management wholeheartedly for attaining market share.
- Implementation of statistical practices on the process reduced rejections and rework in the organization.
- The organization recorded 55% savings in tool bit consumption by adopting a continuous improvement approach.
- Enhancement of training programs after implementing LMPs produces great results. In other words, it can be said that people practices can boost the adoption of newer methodologies if people are properly trained to do so (Singh et al., 2013).
- To reduce breakdown hours, a preventive maintenance policy was effectively employed on the machine shop for improving OEE and productivity.
- Even with a piecemeal approach, LMPs can produce significant results. Lean is a continuous journey toward perfection, and adopting it in daily routines takes less time.
- Looking at the sustainable developments achieved, LMPs enhanced the economic and social performance of the organization.

Managerial implications: The managers played a pivotal role in effective implementation of LMPs on the shop floor of the organization. To enhance the overall

TABLE 10.8

Benefits Obtained by Implementing Lean Practices

S. No.	Type of Waste	Time Period	Frequency	Ease of Removal	Lean Practices	Benefits	Output
1	Removal of unnecessary/ unwanted items from shop floor	Daily (08:00 to 09:00 AM)	High	Easy	5S	Enhancing 5S level on shop floor, defect reduction, scrap value benefits	Delivery rate
2	Overall equipment efficiency	Monitoring on daily basis	High	Difficult	Preventive maintenance (PM)	Improving equipment efficiency, equipment health conditions lead to production of quality product; high efficiency leads to cost effective production, resulting in cost cutting	Firm performance
3	Quality circles/kaizen enhancement	Monthly	Medium	Medium	JIT practices	Implementing more kaizen, cost saving related to product quality, mistake proofing and jig/fixture validation	Productivity
4	Rejections/rework	Daily	High	Easy	SPC/Pareto analysis	Rejections reduced, customer satisfaction and customer delightedness, cost cuttings from defective parts	Quality
5	Raw material planning	Fortnightly	Medium	Difficult	Supplier mgmt.	Reduced inventory leads to buildup of 30+ days inventory and in-time delivery, reduction in waiting loss	Cost
6	Enhancement of training programs	Monthly	Medium	Easy	People practices	Training programs increased from monthly to fortnightly, leads to multiskilling of employees, helping to improve productivity, helps in upgrade of quality tools	Firm performance

firm performance, other tools like process capability index, ERP systems for managing inventory and visual information systems can be used. Others can see from this that using all the LMPs in a short span of time is not viable. They can seek the services that have completely transformed into lean via consultancy firms providing lean services. To enhance sustainability in the organization, the managerial level should focus on enhancing environmental performance to balance the ecosystems on this planet.

REFERENCES

Abdulmalek, F.A. and Rajgopal, J. (2006). Analyzing the benefits of lean manufacturing and value stream mapping via simulation: A process sector case study. *International Journal Production Economics*, 107, 223–236.

Antony, J. (2011). Six Sigma vs Lean: Some perspectives from leading academics and practitioners. *International Journal of Productivity and Performance Management*, 60(2), 185–190.

Bai, C., Satir, A., and Sarkis, J. (2019). Investing in lean manufacturing practices: An environmental and operational perspective. *International Journal of Production Research*, 57(4), 1037–1051.

Belhadi, A., Touriki, F., and Elfezazi, S. (2019). Evaluation of critical success factors (CSFs) to lean implementation in SMEs using AHP: A case study. *International Journal of Lean Six Sigma*, 10(3), 803–829.

Bergenwall, A.L., Chen, C., and White, R.E. (2012). TPS's process design in American automotive plants and its effects on the triple bottom line and sustainability. *International Journal of Production Economics*, 140(1), 374–384.

Caldera, H.T.S., Desha, C., and Dawes, L. (2019). Evaluating the enablers and barriers for successful implementation of sustainable business practice in lean SMEs. *Journal of Cleaner Production*, 218, 575–590.

Chauhan, G. and Chauhan, V. (2019). A phase-wise approach to implement lean manufacturing. *International Journal of Lean Six Sigma*, 10(1), 106–122.

Deshmukh, S.G., Upadhye, N., and Garg, S. (2010). Lean manufacturing for sustainable development. *Global Business and Management Research: An International Journal*, 2(1), 125.

Ghosh, M. (2013). Lean manufacturing performance in Indian manufacturing plants. *Journal of Manufacturing Technology Management*, 24(1), 113–122.

Gupta, S. and Jain, S.K. (2015). An application of 5S concept to organize the workplace at a scientific instruments manufacturing company. *International Journal of Lean Six Sigma*, 6(1), 73–88.

Gurumurthy, A. and Kodali, R. (2010). Design of lean manufacturing systems using value stream mapping with simulation. *Journal of Manufacturing Technology Management*, 22(4), 444–473.

Habidin, N.F., Hibadullah, S.N., Mohd Fuzi, N., Salleh, M.I., and Md Latip, N.A. (2018). Lean manufacturing practices, ISO 14001, and environmental performance in Malaysian automotive suppliers. *International Journal of Management Science and Engineering Management*, 13(1), 45–53.

Iranmanesh, M., Zailani, S., Hyun, S.S., Ali, M.H., and Kim, K. (2019). Impact of lean manufacturing practices on firms' sustainable performance: Lean culture as a moderator. *Sustainability*, 11(4), 1112–1132.

Jain, R.K. and Samrat, A. (2015). A study of quality practices of manufacturing industries in Gujarat. *Procedia-Social and Behavioral Sciences*, 189, 320–334.

Knol, W.H., Slomp, J., Schouteten, R.L., and Lauche, K. (2019). The relative importance of improvement routines for implementing lean practices. *International Journal of Operations & Production Management*, 39(2), 214–237.

Kumar, S.B. and Abuthakeer, S.S. (2012). Productivity enhancement by implementing lean tools and techniques in an automotive industry. *Annals of Faculty Engineering Hunedoara-International Journal of Engineering*, 167–172.

Mehta, K.R., Mehta, D. and Mehta, K.N. (2012). An Exploratory Study on Lean Manufacturing Practices. *Journal of Management and Economics*, 2, 289–299.

Mohanty, R.P., Yadav, O.P., and Jain, R. (2006). Implementation of lean manufacturing principles in auto industry. *Vilakshan XIMB Journal of Management*, 1(1), 1–32.

Naveen, K., Sunil, L., Sanjay, M.K., and Abid, H. (2013). Facilitating lean manufacturing systems implementation: Role of top management. *International journal of Advances in Management and Economics*, 2, 1–9.

Pakdil, F. and Leonard, K.M. (2017). Implementing and sustaining lean processes: The dilemma of societal culture effects. *International Journal of Production Research*, 55(3), 700–717.

Parkash, D. and Kumar, S. (2011). Implementation of lean manufacturing principle in auto industry. *Industrial Engineering Letters*, 1(1), 56–60.

Parry, G.C. and Turner, C.E. (2006). Application of lean visual process management tools. *Production Planning & Control*, 17(1), 77–86.

Piercy, N. and Rich, N. (2015). The relationship between lean operations and sustainable operations. *International Journal of Operations & Production Management*, 35(2), 282–315.

Rajenthirakumar, D. and Thyla, P.R. (2011). Transformation to lean manufacturing by an automotive component manufacturing company. *International Journal of Lean Thinking*, 2, 1–13.

Rameez, H.M. and Inamdar, K.H. (2010). Areas of lean manufacturing for productivity improvement in a manufacturing unit. *World Academy of Science, Engineering and Technology*, 584–587.

Ramnath, V.B., Elanchezhian, C., and Kesvan, R. (2010). Application of Kanban system for implementing lean manufacturing. *Journal of Engineering Research and Studies*, 1(1), 138–151.

Rose, A.M.N., Deros, B.M., Rahman, M.A., and Nordin, N. (2011). Lean manufacturing best practices in SMEs. *Proceedings of the 2011 International Conference on Industrial Engineering and Operations Management*, 2(5), 872–877.

Ruhani, A.R. and Muhammad, A.A. (2012). Production flows analysis through value stream mapping: A lean manufacturing process case study. *Procedia Engineering*, 41, 1727–1734.

Sahoo, T., Banwet, D.K., and Momaya, K. (2011). Strategic technology management in the auto component industry in India. *Journal of Advances in Management Research*, 8(11), 9–29.

Saini, S. and Singh, D. (2018a). Lean practices for consummating competitive priorities in SMEs: A critical review. *International Journal of Business Continuity and Risk Management*, 8(2), 106–123.

Saini, S. and Singh, D. (2018b, 25–26 Oct). Enhancing manufacturing performance through lean perspective: An evidence from smaller enterprise in Northern India. *Proceedings of All India seminar on Micromachining held at, BBSBEC, Fatehgarh sahib*, 87–92.

Saini, S. and Singh, D. (2020a). An evaluation of the status of lean manufacturing practices in SMEs in Northern India. *International Journal of Business Continuity and Risk Management*, 10(4), 330–370.

Saini, S. and Singh, D. (2020b). Impact of implementing lean practices on firm performance: A study of Northern India SMEs. *International Journal of Lean Six Sigma*, 11(6), 1019–1048.

Saini, S. and Singh, D. (2022a). Evaluating the critical success factors for lean implementation in SMEs in northern India using VIKOR approach. *International Journal of Business Excellence* (Inderscience Publications). doi: 10.1504/IJBEX.2020.10038326

Saini, S. and Singh, D. (2022b). Development of a model to assess the impact of lean practices on firm performance in SMEs. *International Journal of Process Management and Benchmarking*, 12(4), 513–542.

Saini, S. and Singh, D. (2022c). Lean manufacturing practices for enhancing firm performance in medium enterprises: a case study from Indian context. *International Journal of Productivity and Quality Management*, 35(3), 352–382.

Sajan, M.P., Shalij, P.R., and Ramesh, A. (2017). Lean manufacturing practices in Indian manufacturing SMEs and their effect on sustainability performance. *Journal of Manufacturing Technology Management*, 28(6), 772–793.

Singh, K. and Ahuja, I. (2015). An evaluation of transfusion of TQM-TPM implementation initiative in an Indian manufacturing industry. *Journal of Quality in Maintenance Engineering*, 21(2), 134–153.

Singh, D., Oberoi, J.S., and Ahuja, I.S. (2011). A survey of literature of conceptual frameworks assessing supply chain flexibility. *International Journal of Applied Engineering Research*, 2(1), 172.

Singh, D., Oberoi, J.S., and Ahuja, I.S. (2013). An empirical examination of barriers to strategic flexibility in Indian manufacturing industries using analytical hierarchy process. *International Journal of Technology, Policy and Management*, 13(4), 313–327.

Sullivan, M.D. and Aken, V.M.E. (2002). Equipment replacement decisions and lean manufacturing. *Robotics and Computer Integrated Manufacturing*, 18, 255–265.

Taj, S. (2008). Lean manufacturing performance in China: assessment of 65 manufacturing plants. *Journal of Manufacturing Technology Management*, 19(2), 217–234.

Thomas, A., Francis, M., John, E., and Davies, A. (2012). Identifying the characteristics for achieving sustainable manufacturing companies. *Journal of Manufacturing Technology Management*, 23(4), 426–440.

Upadhye, N., Desmukh, S.G., and Garg, S. (2010). Lean manufacturing system for medium size enterprises: an Indian Case. *International Journal of Management Science and Engineering Management*, 5(5), 362–375.

Veres, C., Marian, L., Moica, S., and Al-Akel, K. (2018). Case study concerning 5S method impact in an automotive company. *Procedia Manufacturing*, 22, 900–905.

Vienazindiene, M. and Ciarniene, R. (2013). Lean manufacturing implementation and progress measurement. *Economics and Management*, 2, 366–373.

Yadav, V., Jain, R., Mittal, M., Panwar, A., and Sharma, M. (2019). An appraisal on barriers to implement lean in SMEs. *Journal of Manufacturing Technology Management*, 30(1), 195–212.

11 A Systematic Examination of the Barriers to Implementation of Six Sigma in Manufacturing Organizations

Harsimran Singh Sodhi, Doordarshi Singh,
Ishbir Singh, Sarina Lim

CONTENTS

11.1 INTRODUCTION

Six Sigma methodologies are a well-known commercial business enterprise practice currently in use in the manufacturing industries. The Motorola Organization introduced this concept in the mid-80s. "Sigma" is a Greek letter used for the mathematical term "good-sized deviation," which measures the deviation from the average in a specific commercial enterprise approach. "deviation" extra than ordinary outcomes in defective products and services that do not meet consumer needs

DOI: 10.1201/9781003269298-11

[Snee, 2010]. Six Sigma focuses on diminishing variation in any technique applied to decrease imperfections in the cycles, reduces piece of delivering and results, making budgetary investment funds to the least line, blast customer charm [Sodhi, et al., 2019] upgrading the nature of item, excellent degree absconds recognition, and decline them to a negligible number. Four parts predictable with million open doors in a business endeavor. New hypotheses and thoughts were mixed with simple principles and factual methodologies that have been current in extraordinary building hovers for a long time. Six Sigma supplanted the idea of total quality administration (TQM) to turn into the point of convergence of remarkable administration and endeavor greatness for about quarter of a century. It alludes back to the standards, apparatus and techniques used to find the reasons for inefficiency and move away from deformities or mistakes in company strategies by zeroing in on the aspects considered essential to clients [Abdul, et al., 2016]. Six Sigma tunes in to the voice of the customers (VOC) to find their objectives, convert them into specs, and utilize them in planning items or processes [Singh, et al., 2017]. The building blocks were higher organization and board standards to shape the premise of unlimited oversight. The impressive end result drove Motorola stock higher, which cleared the course for Six Sigma. Six Sigma utilizes statically determined device instruments and reduction of imperfections in business venture methods and is vigorously data driven; it focuses on exercises of different sorts that comprise creation (item), backer (transport), and method (process duration) in a company. A Six Sigma certificate with assortment of imperfections for remarkable off focusing is given in Table 11.1.

The attainment of 3.4 defects in line with multiple opportunities (DPMO) [Allen, et al., 2005] suggests the Six Sigma methodology has been observed. A barrier is defined as something that falls outside a purchaser specification. The methodology

TABLE 11.1
Number of Defects for Specified Off-Centering of Process and Quality Levels.

	Sigma/Quality Level					
Off-Centering	3 Sigma	3.5 Sigma	4 Sigma	4.5 Sigma	5 Sigma	6 Sigma
0	2,700	465	63	6.8	0.57	0. 02
0.25 sigma	3,577	666	99	12.8	1.02	0.0063
0.5 sigma	6,440	1,382	236	32	3.4	0.019
0.75 sigma	12,288	3,011	665	88.S	11	0. 1
1.0 sigma	22,832	6,433	1,350	233	32	0.39
1.25 sigma	40,1 1 I	12,20 I	3,000	577	88.5	1
1.5 sigma	66,803	22,800	6,200	1,350	233	3
1.75 sigma	1,05,601	40,100	12,200	3,000	577	11
2.0 sigma	1,58,700	66,800	22,800	6,200	1,300	32

Source: Henderson 2000

may be applied to clients inside (internal) or outside (external) the company and is called define, measure, analyze, improve, and control (DMAIC) [Sodhi, et al., 2019]. Every step is designed to assist an employer in making upgrades in its enterprise methods [Andersson, et al., 2014].

11.2   REVIEW OF LITERATURE

The importance of Six Sigma was assessed on wide electric powered controlled (GE) and desirable results were also found during the implementation [Assarlind, et al., 2012]. Six Sigma guided GE in the sorting out and execution of grouped commitments and the usage of a five-step system to adjust to issues to accomplish at Six Sigma levels. However, Six Sigma ended up being dispatched as a four-improvement system (DMAIC). Eventually, structure degrees were added to underscore the need to have a properly checked assignment, which later became DMAIC.

Undertakings need to reliably attempt to improve themselves. Affiliations must concentrate in at the showcase characteristics of chief cycles or frameworks to see and eliminate absconds which are of number one significance to customers [Snee, 2010]. Six Sigma is a business undertaking that draws on associations to extend their focal points with the goal of streamlining their games, spoofing extraordinary and butchering. It involves innovative thinking to perform, develop and keep up business accomplishment through understanding the necessities of the buyer [Gnoni, et al., 2013].

Setting out on Six Sigma application involves creating top-notch products while fundamentally limiting all internal insufficiencies [Hajmohammad, et al., 2013]. Six Sigma is pertinent for any undertaking that produces product or control for clients. Six Sigma is a huge business improvement action as opposed to handiest a wonderful side intrigue. Little and medium gathering zone should get keep up of demonstrated mechanical business undertaking improvement program like Six Sigma to satisfy the general conflict. SMEs have a sharpness that it's far appropriate only for enormous associations with higher troublesome work and money related assets. There may be additionally a view that Six Sigma execution will add to their cost without a great deal of hypothesized benefits [Sodhi, et al., 2014]. Smaller organizations may moreover have different reservations with respects to Six Sigma use. The issues guiding SMEs in preparing for of Six Sigma ought to be thought of and coordinated to create interest in this progression technique through the SME. The issues and parts that might prove obstacles for the choice of Six Sigma through SMEs ought to be settled and responded to in due order to overcome them [Oberoi, et al., 2008].

The Six Sigma approach originated in 1985 when Motorola began to make its introduction checks genuinely deserts free. In 1988, Motorola got the Malcolm Baldrige National Quality Award, which set the norm for outstanding associations to copy [Naslund, et al., 2017]. Companies are getting a handle on different structures, such as ISO 9000, TQM, and so forth to as-incredible five star. Be that as it may, these structures prohibit the budgetary detail of the endeavors. Thirty years of Six Sigma reveal that if it is followed as it should be, companies can guarantee profits [Myrdal, et al., 2017].

11.3 BARRIERS TO IMPLEMENTATION OF SIX SIGMA FOR SMES

Every organization works to eliminate unwanted barriers in implemented systems in order to enhance productivity and improve the smoothness of the systems. Six Sigma methodologies for have been commonly associated with large companies. Medium-sized companies have moreover started seeing budgetary benefits from this method. They're accomplishing additionally appealing money related speculation assets from Six Sigma responsibilities and higher turn of events. Nonappearance of understanding and getting ready and a bit of the disarrays around Six Sigma has made the SMEs to be vigilant generally the significance of Six Sigma for them. Alongside those there are several certifiable mechanical, definitive and cash related limitations of SMEs that go about as tangles for 6 sigma execution with the benefit of the use of them [Naslund, et al., 2017]. Each of those obstacles is considered in the following.

1. Lack of leadership commitment
2. Incomplete understanding of Six Sigma methodologies
3. Poor execution
4. Lack of resources
5. Internal resistance
6. Lack of leadership from top executives

11.3.1 LACK OF LEADERSHIP COMMITMENT

It has been found that an organization's concern in adopting Six Sigma comes when management chooses which work force should implement it. Utilizing whoever is accessible in region of committing zenith comprehension to 6 sigma task endeavors areas the undertaking on questionable balance and diminishes the chances of its accomplishment. The Six Sigma task requires pioneers willing to commit resources of time, capacities, and money to the endeavor [Myrdal, et al., 2017].

11.3.2 INCOMPLETE UNDERSTANDING OF SIX SIGMA METHODOLOGIES

Many times it has been observed that management and workers do not have a proper understanding regarding the Six Sigma methodologies at the time of its implementation. Therefore, it is always recommended that proper training programs should be carried out in order to ensure a complete understanding of Six Sigma methodologies at the every level of the organization. Of their energy to hoard the upsides of passing on the Six Sigma methodology, a few social occasions flood in sooner than they've an undertaking hold close of what a victory Six Sigma execution calls for. This could upward push up while companies put into sway Six Sigma truly to set aside with the restriction, or to impact financial specialists with the guide of being good for use non-thwart strategy improvement stating in office documentation. Ventures that regard Six Sigma as a panacea or implement it without the sources it requires are inviting dissatisfaction. Many institutes and coaching centers are also providing certifications at various levels after training the workforce at various levels and companies can overcome this obstruction through submitting truth be told to the contraption and using and helping Six Sigma aces with checking that the affiliation is sending the procedure and no longer actually the utilization of the expressing. The ones experts other than keep up the undertaking zeroed in on focus exercises wherein they may have the alternative to make the most separation, no longer essentially at the straightforward changes and the low striking natural item.

11.3.3 POOR EXECUTION

Proper execution of Six Sigma methodologies also an important aspect for the implementation of this technique. In fact, even underneath the master coordinating of experience champions and handle dim belts, Six Sigma astonishing headway assignments can drift off track if they will be by and by not charmingly wrapped up [Kocak, et al., 2017]. Negative execution happens even as technique overhauls do not agreed with the business's needs, even though the undertaking relies upon responsively understanding issues in tendency to meet crucial goals or while the improvement mission rehearses inside the yield of the contraption as opposed to the wellsprings of data. Companies should always remember that Six Sigma techniques aren't guaranteed to work in a vacuum anyway that they craftsmanship remarkable on comparative time as agreed with the needs and needs of the undertaking, they may be significantly more inclined to live on course [Hajmohammad, et al., 2013]. Associations that discover they'll be finished getting the benefit salary or money related financial budgetary hold supports they foreseen from using Six Sigma approach aren't disturbed an immediate aftereffect of the truth the technique is vain. The best likely wellspring of their wretchedness is that the duties need fantastic organization and are controlled inefficiently. While authority is devoted to utilizing the Six Sigma procedure, apportions top blessings to undertaking social affairs, gets the undertaking through a real choice and examination system, and gives the upheld property, the odds of Six Sigma achievement impact radically [Snee, 2010].

11.3.4 LACK OF RESOURCES

Lack of resources is one of the major barriers in the implementation of Six Sigma methodologies. A Six Sigma strategy would possibly eat really a few significant business re-resources like cash related property, human possessions, time, etc. For usage of Six Sigma, affiliations essentially require composed and showed work for imagining and walking Six Sigma commitments. Enormous planning for an aspect of the picked successful people in the business is fundamental to offer any essential Six Sigma impacts. A Six Sigma experience pioneer is known as a dull belt who works for full time on ventures [Gnoni, et al., 2013]. The undertaking alliance individuals are known as Green belts (GB) who regularly canvases part of their time on experiences. Dark Belts (BBs) and Green Belts (GBs) ought to be tirelessly given express trainings to manage their positions reasonably. Hold close dull belts are experienced BBs who have chipped away at different duties. Mbbs direct the BBS and cutoff of the status work will be their dedication. The rate attracted with training the extraordinary degree chiefs changed into a tangle for the humbler companies for a certifiable extended length. Studies with basic Six Sigma affiliations show that the cash related compensation outperform the speculation. Thusly, immense relationship with capacity to address Six Sigma attempts gathered royal growths in some dark time later on for. Gen iii Six Sigma has changed this situation extensively. The instructing and framework expected to get ideal outcomes from a gen iii mission are a game plan less. Gen iii has brought the opportunity of the white belt Six Sigma practitioner who awards utilization of Six Sigma in craftsmanship cells or proportionate set-tolls. White belts raise benefits with the guide of utilizing Six Sigma to issues that would not legitimize the time and thought of a Six Sigma dull belt. SMEs have begun seeing that Six Sigma might be overcome relationship of any range and wide benesuits might be developed.

It does no longer by and large make essential for the relationship to move for silly changes in their creation or director plan with goliath financing for 6 sigma usage. For cutoff of the upgrades the certified need is to explore the property of goofs and design and deal with the exercises to put off them.

11.3.5 INTERNAL RESISTANCE

Internal resistance is a major challenge that has been faced by the manufacturing organization during the implementation of Six Sigma methodologies in the industrial organizations. Any change in any alliance will dependably meet with a specific extent of obstruction and mindfulness. Without the certification by workers, Six Sigma measure improvement will indeed come up short. In any association, little or huge, representatives will get acclimated with the cycle they are utilizing for quite a while. Protection from change this is ordinary and it ought to be anticipated by the Six Sigma get-togethers. For delegates the principal procedure for showing limitation is by repelling the new applications or mechanical congregations and overlooking the new cycle. Be that as it may, smashing the impediment of individuals at all levels is extraordinarily major for effective utilization of Six Sigma. However, if the focal points are clear, there will be protection from change. Companies likewise may get skeptical about the central focuses and along these lines may waver to help the Six Sigma works out. In the event that there is solid affirmation demonstrating the accomplishment of the execution—the estimation of the cycle gets more straightforward. Correspondence is the best way to deal with manage such an issues. By presenting about the positive aftereffects of Six Sig mother, pioneers can empower sureness of individuals and enroll their help for the change cycle. For this the pioneers ought to have enough events of occurrences of beating distress close by before revealing the new cycle. Deliveries, messages or conversations will support consequently. Executing measure change enough requires limits that are less in any Six Sigma instructional handout. Usage of Six Sigma requires some extraordinary choice from specific extent of limits. The hidden development to beneficial Six Sigma use is compassion. To spread it out just, everyone will ask themselves, "In what limit may this preferred position me?" This solicitation is a key bit of human sense. Right when every individual from the social event handles by what strategy may this preferred position him, he/she can consolidate wholeheartedly and offer better. One of the basic wellsprings of obstacle is the conviction that Six Sigma will affect decrease in workforce. Focus of Six Sigma isn't only enlargement in capacity yet also increment in sensibility. Six Sigma goes after for progression of abundancy will incite improvement of business and not stifling. Hence, it is the commitment of the pioneers to prepare the individuals of the relationship in such manner so that as opposed to confining the utilization of Six Sigma they begin developing their full venture. In the event that representatives are not orchestrated appropriately on the new cycle, they may negate it imagining that it is hard to finish. Arranging about the need and the systems of the Six Sigma execution may help individuals with holding changes better. Preparing ought to change them into the guaranteed 'change agents'. A proposition/input system from all the concerned individuals for measure update and tenacious improvement should be set up. This aides in giving an assumption of possession to the workers and any following improvement because of their duty will request that they partake in the process intentionally. Opposition could in like way be utilized as a contraption to improve the program and guard against foolish utilization of Six Sigma. The change ace ought to inquire as to why the snag exists, and what can be gotten from it to make the program able to all. Indeed, once in a while Six Sigma ends up being answer for the issues of individuals, if the work-ers are fomented, obviously uninvolved, confined or trading off in the unavoidable climate, by giving them undeniably more prominent chance to show their inven-tiveness which they were missing prior. In a Six Sigma experience the guideline drivers of the issues ought to be settled, evaluated and dismantled by the task accomplices utilizing their experience and innovativeness. During the improvement stage accomplices should conceptualize thoughts to make improvement in execu-tion with a definitive target that the sigma level of the cycle moves to 6 sigma level. Social affairs with astounding insights basically can improve sigma execution. The environment in SMEs is more helpful for these happenings be-clarification behind the smaller idea of the association and better coordination between the association and the agents.

11.3.6 Lack of Leadership from Top Executives

Because of weak position and nonappearance of commitment from senior associa-tion, Six Sigma is seen by explicit relationship as only a passing association winning style and subsequently may not show excitement for it. Affiliations require innovative relationship at all degrees of the relationship for advancement and development. It is improvement, not words that make the triumphs. For real implementation of Six Sigma an association should have visionary top affiliation that examines the cycles what's more perceives how to make a collusion that cooperates. The top affiliation should keep up, sponsorship and give re-sources with an authoritative objective that Six Sigma changes into an engaging manager to achieving the business fights of the association.

Six Sigma has been profitable generally considering the way that the results got by its execution have pulled in the management. A part of the declarations from specialists concerning implementation of Six Sigma by SMEs are recorded underneath: All affiliations goliath and little, share diverse major features and issues. Tremendous affiliations, because of scale, may obtain higher budgetary preferences thinking about a given advancement, regardless this should not be taken to suggest that little affiliations would not benefit colossally from its utilization. Joseph De Feo, CEO of Juran Institute, USA. "Six Sigma is fabulously appropriate for humbler affiliations too. The Six Sigma structure works superbly in billion dollar corporations in like manner as $50 million covertly held affiliations. Really, it has been our experience that the results are normally snappier and more observable in humbler affiliations" Dr Matthew Hu, Vice President of Technology and Innovation, ASI, USA. Hence, relationship of little affiliations can't give pardons like Six Sigma sending in their affiliations would not be practicable or not significant. Past undertakings at improvement may have without a doubt failed in a relationship considering the nonattendance of the heads keep up, no structure related to its implementation and nonappearance of the principle get-together of the declaration of progress. Definitely when an affiliation adequately gets included and reinforces something it will work. That is the significant partition between Six Sigma and the other quality exercises. A Six Sigma affiliation will be proactive rather than open.

11.4   METHODOLOGY ADOPTED

Extensive review of literature put together absolutely generally with respect to different waste control methods has been explored. From the writing it is been found that there some of impediments in the usage of waste control systems in the assembling organizations. A poll on a licker size of five come to be composed to get to the degree of impediments in assembling organizations. Further, this poll end up dispatched to various analysts, academicians and undertaking specialists with the end goal of a pilot review and approval. After you have the significant information sources, the poll have become finished. Step through advance procedure followed for this have an investigate is outlined in figure 11.1.

The Manufacturing SMEs have been recognized from at some phase in India for realities arrangement. Additionally, the concluded poll adjusted into sent to 600 and fifty SMEs to catch the voice of mechanical organizations concerning the accomplishment of did squander oversee strategy. Out of this, 100 and thirty had reacted, and a response pace of 21%. After end of unusable reactions prompted 100 and 21 reactions at last for the what's more appraisal works of art. This contrasts pleasantly and the response expenses for research in activities oversee (Oberoi, Et Al., 2008; Singh, Et Al., 2013) it's been resolved that differing SMEs during India are the utilization of one-of-a-type squander control systems like lean creation, Six Sigma, 5S, TPM, LSS, etc. A spreadsheet have gotten composed for gathering the voice of various SMEs with respect to the accomplishment of applied waste.

The fame of the examination is on little and medium-scale producing firms across India. The data amassed for the current examination comprises of sixty one% reactions from little scope and 39% reactions from medium scale organizations and appeared in Figure 11.2.

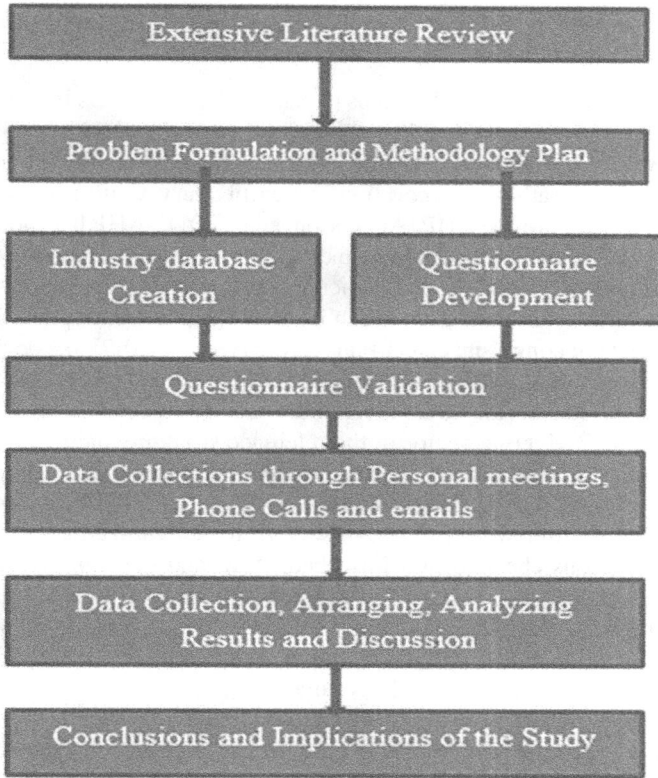

FIGURE 11.1   Methodology adopted for the study.

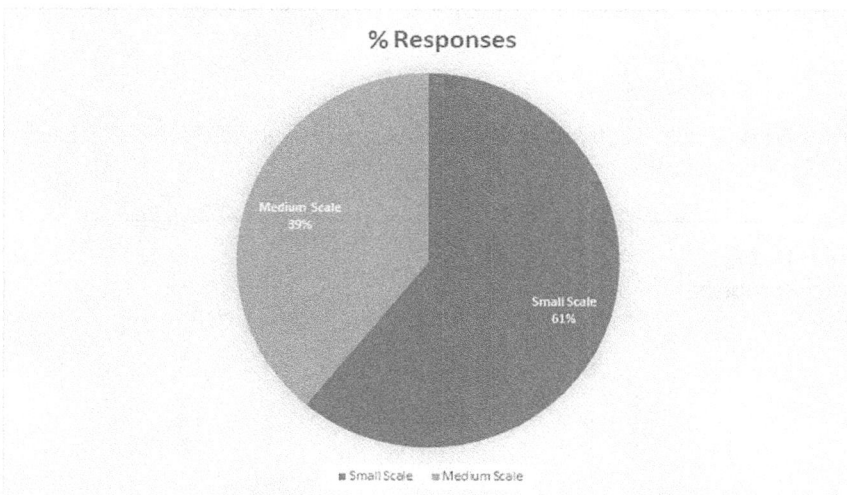

FIGURE 11.2   Breakdown of responses.

11.5   ANALYTICAL HIERARCHY SYSTEM FOR ASSESSING BARRIERS TO WASTE CONTROL STRATEGIES

The AHP is a system to survey the heaps of criteria. The AHP has been used by and large in an assurance of bewildered need making issues related with crucial organizing of various leveled assets, the evaluation of key alternatives[Sodhi, et al., 2014], or the guard of late gathering innovation. An earlier survey furnished in excess of hundred related packs of the AHP [Andersson, et al., 2014]. AHP has moreover been used remarkably in exhibiting and advancing and promoting, support, mentoring, public technique, budgetary angles, medicine, and sports practices sports. AHP has been applied in a spread of codecs which include: the plan contraption for colossal degree structures or composite extent scales [Kocak, et al., 2017], the device for pair wise evaluation inside the utility of produced neural organizations, or the basic kind of choice oversee structures. As an accessible strategy, the AHP approach has been used to check the weighting vector in the reference structure and are searching for reference course in a seen clever contraption, and to discover point coefficient and limit regards in more than one-desire lp burdens. Version of TPM tendency creation using the AHP. A framework that uses the AHP for legitimization of TPM use in progress associations changed into furnished. AHP and taguchi adversity feature to advancement a tendency variation to help tendency makers with need of the correct provider for the redistributing purposes. It come to be inferred that paying little regard to the way that re-appropriating gives tremendous preferences for the association, it includes with it severe hazards as suitably.

The degrees of obstructions in the execution of Six Sigma in progress affiliations were researched the utilization of AHP frameworks. Character of the essential credits (limits for this circumstance) for AHP requires an outrageous assessment of the trouble. For cutting edge have a look at, the selection of qualities has been settled through composing survey and discussions which have been held with specialists all through current visits and with academicians running in the undefined region.

11.5.1   DECISION MATRIX

The resulting weights are based on the principal eigenvector of the decision matrix in Table 11.2. Calculated eigen value is 7.612.

TABLE 11.2
Decision Matrix

	1	2	3	4	5	6
1	1	6	4	9	3	4
2	0.17	1	1	6	1	3
3	0.25	1	1	9	7	8
4	0.11	0.17	0.11	1	3	6
5	0.33	1	0.14	0.33	1	2
6	0.25	0.33	0.12	0.17	0.5	1

TABLE 11.3
Pairwise Comparison

S. No	List of Barriers	Priority	Rank	(+)	(-)
1	Lack of leadership commitment	0.433	1	0.279	0.279
2	Incomplete understanding of six sigma method	0.143	3	0.118	0.118
3	Poor execution	0.25	2	0.168	0.168
4	Lack of resources	0.076	4	0.065	0.065
5	Internal resistance	0.063	5	0.041	0.041
6	Lack of leadership from top executives	0.035	6	0.026	0.026

11.5.2 PRIORITIES

These are the resulting weights for the criteria based on your pairwise comparisons as shown in the Table 11.3.

11.6 CONCLUSIONS

A thorough investigation of various factors responsible for the barriers for the implementation of Six Sigma in the manufacturing organizations has been done. From the AHP calculations it has been observed that Lack of Leadership Commitment with priority 0.433 is ranked as first among all factors. Afterwards Poor Execution is ranked at second number with a priority of 0.25. Incomplete Understanding of Six Sigma Method with a priority of 0.143and it's ranked at number three among other barriers for the implementation of Six Sigma in the manufacturing organizations. Lack of resources is again a serious barrier in the implementation of Six Sigma and it's ranked at number four with a priority of 0.076. Further it has been noticed that internal resistance in the organization towards the implementation of Six Sigma is another major factor with a priority of 0.063 and it's ranked as fifth among all considered factors. Lack of leadership from top executives is a major barrier identified from the reviewed literature is ranked at no six with a priority of 0.035.

REFERENCES

Abdul, S., Albliwi, S., Antony, J. and Vander, T. (2016), "Critical failure factors of lean Six Sigma: a systematic literature review", *International Journal of Quality and Reliability Management*, Vol. 31 No. 9, pp. 1012–1030.

Albliwi, S., Antony, J., Abdul, S. and Vander Wiele, T. (2014), "Critical failure factors of lean Six Sigma: a systematic literature review", *International Journal of Quality and Reliability Management*, Vol. 31 No. 9, pp. 1012–1030.

Allen, D., Gutowski, T., Murphy, C., Bauer, D., Bras, B. and Wolff, E. (2005), "Environmentally benign manufacturing: observations from Japan, Europe, and the United States", *Journal of Clean Production*, Vol. 13 No. 1, pp. 1–17.

Andersson, R., Hilletofth, P., Manfredsson, P. and Hilmola, O. (2014), "Lean Six Sigma strategy in telecom manufacturing", *Industrial Management and Data Systems*, Vol. 114 No. 6, pp. 904–921.

Assarlind, M., Gremyr, I. and Backman, K. (2012), "Multi-faceted views on a lean Six Sigma application", *International Journal of Quality and Reliability Management*, Vol. 29 No. 1, pp. 21–30.

Gnoni, M.G., Andriulo, S., Maggio, G. and Nardone, P. (2013), "Lean occupational safety: an application for a Near-miss Management System design", *Journal of Safety Science*, Vol. 53 No. 2, pp. 96–104.

Hajmohammad, S., Vachon, S., Klassen, D. and Gavronski, I. (2013), "Lean management and supply management: their role in green practices and performance", *Journal of Clean Production*, Vol. 39 No. 2, pp. 312–320.

Kocak, A., Carsrud, A. and Oflazoglu, S. (2017), "Market, entrepreneurial, and technology orientations: impact on innovation and firm performance", *Management Decision*, Vol. 55 No. 2, pp. 248–270.

Myrdal, L., Shrivastava, A. and Dinesh, S. (2017), "Critical success factors for lean Six Sigma in SMEs (small and medium enterprises)", *The TQM Journal*, Vol. 28 No. 4, pp. 613–635.

Naslund, D., Vaaler, P.M. and Devers, C. (2017), "Same as it ever was: the search for evidence of increasing hyper-competition", *Strategic Management Journal*, Vol. 24 No. 3, pp. 261–278.

Oberoi, J.S., Khamba, J.S. and Sushil, K.R. (2008), "An empirical examination of advanced manufacturing technology and sourcing practices in developing manufacturing flexibilities", *International Journal of Services and Operations Management*, Vol. 4 No. 6, pp. 652–671.

Singh, D., Singh, J. and Ahuja, I.P.S. (2013), "An empirical investigation of dynamic capabilities in managing strategic flexibility in manufacturing organizations", *Management Decision*, Vol. 51 No. 7, pp. 1442–1461.

Singh, S., Singh, B. and Khanduja, D. (2017), "Synthesizing", *International Journal of Entrepreneurship and Innovation Management*, Vol. 19 No. 3, pp. 256–283.

Snee, R. (2010), "Lean Six Sigma—Getting better all the time", *International Journal of Lean Six Sigma*, Vol. 10 No. 1, pp. 9–29.

Sodhi, H.S., Singh, D. and Singh, B.J. (2019), "Developing a lean Six Sigma conceptual model and its implementation: a case study", *Industrial Engineering Journal*, Vol. 12 No. 10, pp. 1–19.

Sodhi, H.S., Singh, G. and Mangat, H.S. (2014), "Optimization of end milling process for D2 (die steel) by using response surface methodology", *Journal of Production Engineering*, Vol. 17 No. 2, pp. 73–78.

12 Modeling of Carburization of Steels
A Sustainable Approach

Ravi Shankar Prasad

CONTENTS

12.1 INTRODUCTION

Carburization is a commonly used case hardening process that enhances functional properties of gears, bearings, shafts, and so on without sacrificing mechanical properties at its core (Banerjee, 2018; Hutchings & Shipway, 2017). The different phases (ferrite, pearlite, bainite, and martensite) that are transformed through heat treatment are affected by cooling rate and chemical composition. The fractional quantity and appearance of the microconstituents that are transformed through heat treatment determine their mechanical properties. Researchers have worked to obtain micrographs after heat treatment of steels and published them in the form of phase and transformation diagrams (Vander Voort, 1991). The process of obtaining this data is uneconomical in terms of time and cost, and therefore attempts have been made (Hillert & Höglund, 2004; Teixeira et al., 2021) to calculate the transformation of phases in various alloy steels after heat treatment. The carburization of steel is a process in which the carbon concentration gradient is set from the surface to the case depth, and the cooling rate at different nodes from the surface to core varies. Therefore, the cooling rate and chemical composition that effect phase transformation through heat treatment can be predicted.

DOI: 10.1201/9781003269298-12

Here, the development of theoretical model was done using a computational tool (ANSYS). Curves with different cooling rates were generated for quenched steel samples. Curves with different cooling rates were generated when the samples were cooled through various quenching and normalizing processes. The curves obtained in this way were placed over the published time temperature transformation (TTT) diagrams. The simulation of phases transformed is affected by supercooling (ΔT) and time. It was assumed that the exponent on time $n = p/t^m$ is a variable in the Johnson-Mehl equation. Micrographical analysis and microhardness tests were done to support the results obtained through simulation work. Further, the theoretically obtained result was compared with the experimentally observed microstructures.

12.2 THEORETICAL MODELING

Theoretical modeling includes (1) calculation of concentration gradient after carburization, (2) ANSYS (FEM package) for generating cooling curves, and (3) simulation of micrograph in heat-treated steel.

12.2.1 CALCULATION OF CONCENTRATION GRADIENT AFTER CARBURIZATION

The equation obtained as a solution to Fick's second law is (Niederschlags, 1851)

$$\left[\frac{(C_X - C_S)}{(C_0 - C_S)}\right] = \mathrm{erf}\left[\frac{x}{2 \times \sqrt{D \times t}}\right] \qquad \text{(Eq 1)}$$

For the given boundary conditions:

At $t = 0$, $C = C_0$, and x lies between zero and infinity.
At $x = 0$, $C = C_s$, and t lies between zero and infinity.

Where $C(x, 0) = C_0$, $C(0, t) = C_s$ and C_x is the percent of carbon at distance x from the surface, D = coefficient of diffusion of carbon in austenitic phase iron, t = time = 10 hours.
The coefficient of diffusion of carbon in austenitic phase is

$$D_{C \, \mathrm{in} \, \gamma-\mathrm{iron}} = D_0 \times e^{(-Q/RT)} = 1.066 \times 10^{-11} \mathrm{m}^2/\sec \qquad \text{(Eq 2)}$$

where $D_0 = 0.7 \times -10^{-4}$ m²/s (a constant), $Q = 157 \times 10^3$ J/mol (activation energy), T = temperature of carburization = 930 °C, R = universal gas constant = 8.314 J/mol/°K. The percentage of carbon at different nodes was calculated using Fick's diffusion (Eq. 1).

12.2.2   ANSYS (FEM Package) for Generating Cooling Curves

The assumptions made to develop the model were:

1. The length to diameter ratio (l/d) > 5, and therefore the heat transfer across the section was considered.
2. The transfer of heat occurred across the radial direction only, and two-dimensional analysis was done using the FEA package (ANSYS).
3. The model constituted a solid circle as core and annuluses to accommodate the case depth, since after carburization, the carbon content varied from surface to the case depth. Nodes G, F, E, D, and C are located at the periphery of circle at each annulus and core. Within the core, including the center, two additional points (B, A) were considered. The percent of carbon at each node was calculated using Eq. 1, and the generated cross-section is as shown in Figure 12.1.
4. The effective film coefficient was considered a constant (Bates, 1988; Buczek & Telejko, 2013; Prediction et al., 2009).
5. The effect of temperature and chemical composition on thermal conductivity was considered through following equation (Sinha et al., 2007)

$$K = a - b \times W + c \times W^2 \qquad \text{(Eq 3)}$$

where K = thermal conductivity (in W/m-°C), $a = 76.8 - 6.68 \times 10^{-2}\,(T)$, $b = 34.2 -9.9 \times 10^{-2}\,(T) + 0.815 \times 10^{-4}\,(T)^2$ $c = 9.31 - 3.96 \times 10^{-2}\,(T) + 0.418 \times 10^{-4}\,(T)^2$, W = percent of elements added in grams (Maher et al., 2014).

For the sample of 20.32 mm, the calculated position of nodes, with their distance from surface to core, were [{G(0 mm) at 0.8%C}, {F(0.31 mm) at 0.6%C}, {E(0.559 mm) at 0.5%C}, {D(1.239 mm) at 0.3%C}, {C(3.469 mm) at 0.2%C}, {B(5.96 mm) at core}, {A(10.16 mm) at core}]. For the sample of 15.6 mm, the calculated position of nodes, with their distance from surface to core, were [{G(0 mm) at 0.8%C}, {F(0.31 mm) at 0.6%C}, {E(0.559 mm) at 0.5%C}, {D(1.239 mm) at 0.3%C}, {C(3.469 mm) at 0.2%C}, {B(5.06 mm) at core}, {A(7.8 mm) at core}]. The effective film coefficient for conventional oil at temperature 65 °C and circulation speed 0.51 m/s was 3000 W/m². °K. The effective film coefficient for water at temperature 32 °C and circulation speed 0.51 m/s was 12,000 W/m². °K. The effective film coefficient for air at temperature 27 °C was 300 W/m². °K.

Specific enthalpy and thermal conductivity values are given in Table 12.1 (Sinha et al., 2007), and curves with different cooling rates, generated through ANSYS, are shown in Figure 12.2.

12.2.3   Simulation of Micrograph in Heat-Treated Steel

The obtained curves were plotted on the TTT diagram where transformation start and finish indicate 0.01 and 0.99 fraction transformation, respectively.

TABLE 12.1
Calculated Value of Thermal Conductivity and Specific Enthalpy

Temp, (°K)	0.2 Percent Carbon		0.3 Percent Carbon		0.5 Percent Carbon		0.6 Percent Carbon		0.8 Percent Carbon	
	K	I	K	I	K	I	K	I	K	I
673	43	216040	42.27	214360	40.81	214365	40.09	216040	38.65	219390
773	38.57	280520	38.05	279470	35.55	279990	36.5	281560	35.46	284700
873	33.34	354200	33.05	351690	32.49	351065	32.23	352950	31.75	356720
973	27.32	439610	27.25	432280	27.25	427155	27.28	429350	27.52	433750
1073	20.51	562710	20.66	550980	21.28	547630	21.66	556000	22.76	572750
1203	10.46	651760	10.92	630340	12.44	620610	13.33	632300	15.59	655680

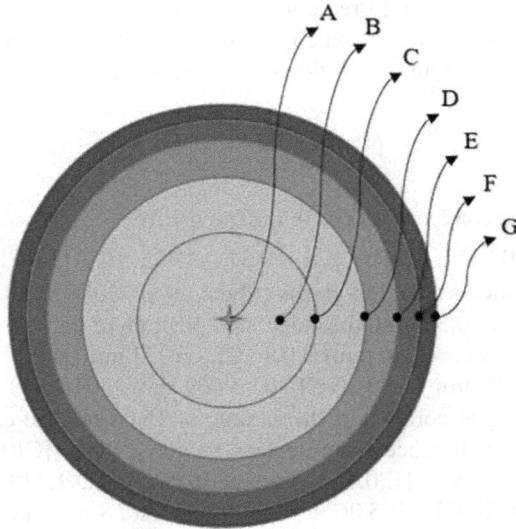

FIGURE 12.1 Distance of nodes from surface for both samples.

12.2.3.1 Calculation of Ferrite Transformation

It was assumed that the nucleation (I) in the austenitic phase takes place randomly and at a constant rate, whereas ferrite transforms at a constant rate (U) as spheres. Until the particles impinge, isothermal ferrite transformation (x) takes place. The transformation is time (t) dependent and is given by the Johnson-Mehl equation (Sinha et al., 2007),

$$x = 1 - \exp\left\{-(\pi/3) \times I \times U^3 \times t^4\right\}$$ (Eq. 4)

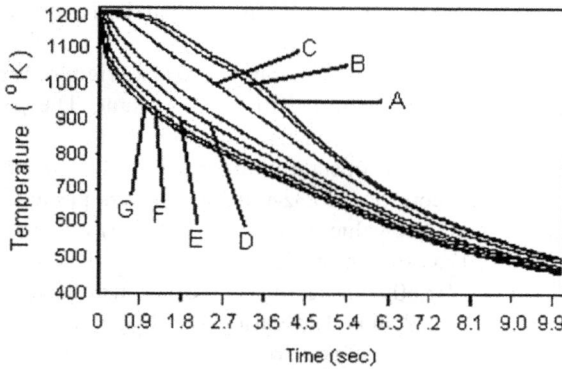

FIGURE 12.2A   Curves with different cooling rates for quenched sample (20.32 mm) (Sinha et al., 2007).

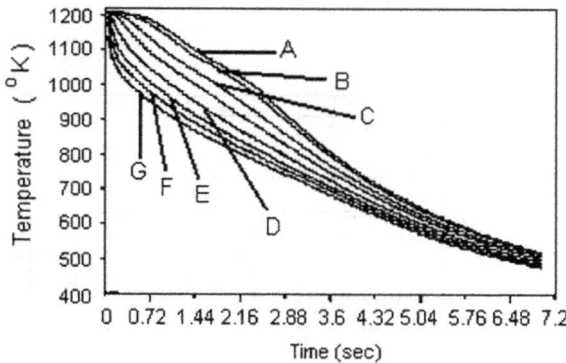

FIGURE 12.2B   Curves with different cooling rates for quenched sample (15.6 mm) (Sinha et al., 2007).

The Johnson-Mehl equation was extended by considering the degree of super cooling (ΔT). Therefore:

$$\ln\left(1-x\right) = -A \times t^4 \times \exp\left[-B / \left\{\left(\Delta T\right)^2 \times T\right\}\right] \qquad \text{(Eq. 5)}$$

where x = fraction of ferrite transformation in time t and temperature T, T_{Fe} = ferrite transformation start temperature, and $\Delta T = (T_{Fe} - T)$.

Eq. 5 was obtained using the equation:

$$I \propto \exp\left[-\left(\Delta f^* + \Delta H_d\right) / RT\right] \qquad \text{(Eq 6)}$$

$$\Delta f^* \propto 1 / \left(\Delta T\right)^2 \qquad \text{(Eq 7)}$$

where $\Delta f^* =$ nucleation barrier, $\Delta H_d =$ the activation barrier for diffusion at interface, and $R =$ gas constant.

For higher temperatures ($T \geq$ nose temperature), ΔH_d was ignored and $\Delta f^* \gg \Delta H_d$. Since U depends on ΔH_d, it is assumed to be a constant. The parametric values affecting A in Eq. 5 are considered constant and independent of percent of carbon; therefore, A can be considered a constant. Similarly B is also considered a constant. The curve obtained at the core of a 20.32-mm sample was plotted on the TTT diagram of 0.2% carbon steel. The value of A and B (in Eq. 5) was calculated using the value at points $a1$ and $b1$, as shown in Figure 12.3. The calculated values of A and B were 0.008987 and 6,502,600, respectively. Further, using Eq. 5, the percent of ferrite transformation was calculated at various nodes across the cross-section of both specimens. The ferrite transformation continued until point $c1$, as shown in Figure 12.3. From point $c1$, pearlitic transformation started, with an assumption that no further ferrite transformation took place.

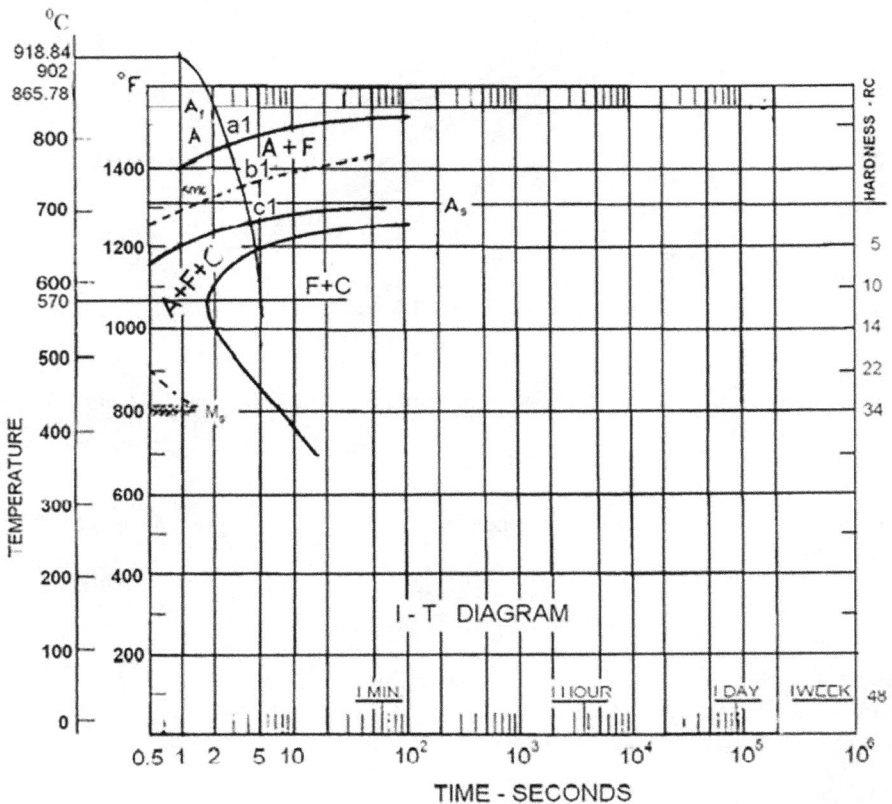

FIGURE 12.3 Generated curve at core of 20.32-mm sample on TTT diagram of 0.2% carbon steel (Vander Voort, 1991).

12.2.3.2 Calculating Pearlite Transformation

The generated curves intersect the pearlite-finish curve, and the fractional pearlitic transformation calculation was done using (1 – ferrite transformation). It was assumed that no martensite transformation took place. The transformation was assumed to be isothermal pearlite transformation, since it was a short temperature range. Also, the TTT and continuous cooling transformation curves show nearly the same values at the nose. Therefore, the cooling curve intersecting the line between the pearlite-start nose and pearlite-finish nose determines the percentage of pearlite transformed. The different cooling curve intersects various other locations and indicates a different fraction of phase transformation. The equation obtained to calculate the amount of pearlite transformation is the following:

$$F = 1 - exp\left(-K \times t^n\right) \qquad \text{(Eq. 8)}$$

where F represents the fractional transformation of pearlitic phase after time t. Eq. 8 is similar to the Johnson-Mehl and Avrami (Sinha et al., 2007) equations with a modification that n varies with time. Since the phase transformation rate retards with time, it was assumed that, $n = p/t^m$. Ignoring higher-order terms, Eq. 8 can be rewritten as,

$$F = K_1 \times t^n$$

Or
$$\ln F = \ln K_1 + n \ln t,$$

Or
$$\ln F = \ln K_1 + \left(p / t^m\right) \times \ln t$$

The data for calculations are obtained from Figure 12.4. The nose temperature (T_N) was 538 °C, and the time required for the phase transformations of 1% and 99% was 0.814 and 5.45 seconds, respectively. Trials were done to select the value of m as 0.66. Similarly, for a node at the surface with 0.8% carbon, the fraction transformation can be given by $F = 1 - exp\left(-0.03935 \times t^n\right)$, where $n = 5.8239/t^{0.66}$. After a duration of 2.39 seconds and at nose temperature of 538 °C, the transformation is found to be 0.495. Since the values of K_1, p, and m vary with carbon content, using different transformation diagrams, other pearlitic transformation equations can also be generated.

12.2.3.3 Calculating Martensitic Transformation

The transformation of martensitic phase can be calculated using the equation,

Transformation of martensite = [1 – (ferrite transformed
+ pearlite transformed)] (Eq. 9)

The theoretically calculated phase transformations of ferrite, pearlite, and martensite from austenite are as plotted and shown in Figure 12.5.

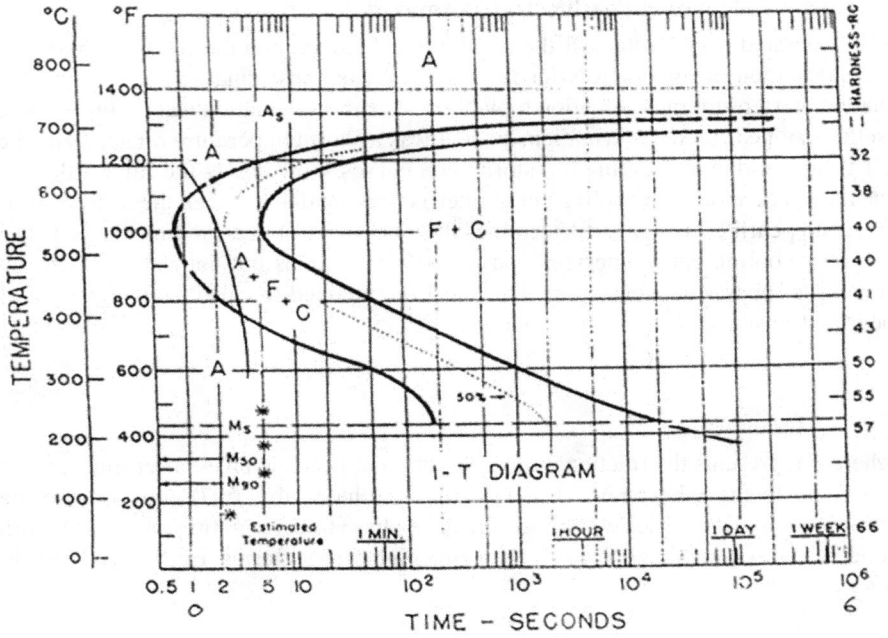

FIGURE 12.4 Generated curve at surface of 15.6-mm sample on TTT diagram of 0.8% carbon steel (Vander Voort, 1991).

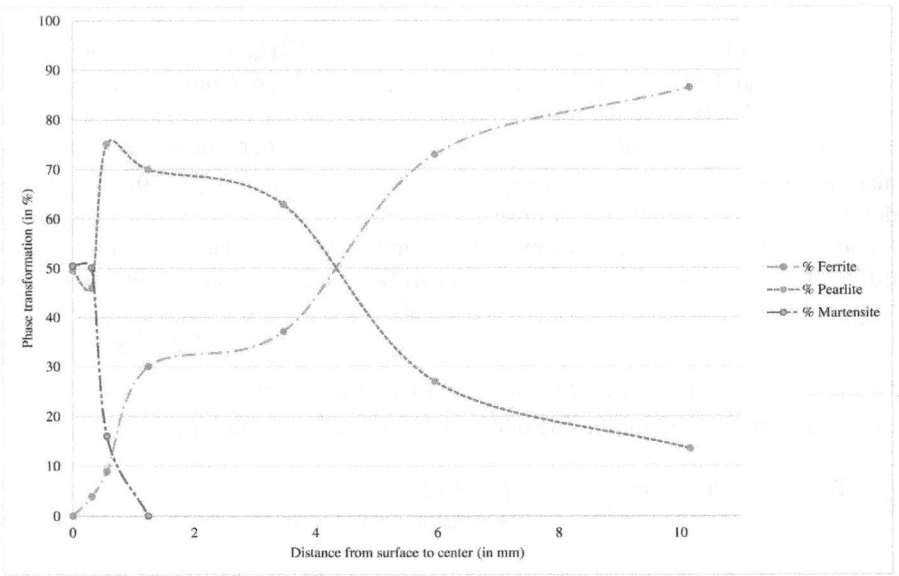

FIGURE 12.5 Transformed phases through calculation.

12.3 EXPERIMENTAL PROCEDURE

The material selected for study was 0.2% carbon steel and 0.16% carbon steel of diameter 20.32 and 15.6 mm, respectively. The selection of ratio of length to diameter was 8, in order to consider the transfer of heat across the radial direction only. Both samples were tested, and the chemical composition of the 20.32- and 15.6-mm diameter samples was C 0.16%, Si 0.023%, Mn 0.68% and C 0.2%, Si 0.22%, Mn 0.82%, respectively.

The samples were annealed to study un-carburized material. A metallic jar was fabricated, and samples were kept with the barium and carbon mixture (in 1:4 ratio). At 670 °C, the jar was placed inside the muffle furnace. The temperature was raised to 930 °C and held for 10 hours as soaking time. After 10 hours, the samples were taken out and quenched in oil for 2 minutes. The samples for study were taken from the middle of the bar. Microstructural study across the cross-section of the sample was done using an image analyzer. Microhardness tests (at 400x) were conducted in order to confirm the phase transformation results obtained through the image analyzer. The graphical point count method was also used to verify the results obtained through the image analyzer and microhardness tests. After carburization, the samples were annealed at 900 °C for 20 minutes in order to get better micrographs for further analysis.

12.4 RESULTS AND DISCUSSION

Microstructural analysis after annealing of the steel samples at 900 °C indicated low-carbon steel. The carbon concentration obtained experimentally at different nodes along the case depth was very close to the theoretically obtained values using Fick's diffusion (Eq. 1). The experimentally observed depth of the carburized case was also very close to the theoretically calculated value, which is 3.469 mm. It was observed that annealing for 20 minutes did not have any significant effect on the carbon concentration from the surface to the case depth. Therefore, Fick's diffusion equation holds to predict the soluble carbon from the surface to the case depth in carburized steel. The theoretically obtained carbon at different nodal points was used for further analysis. The microstructure at the surface (point G) indicated that the austenitic phase transformed to nearly 50% pearlite and 50% martensite. Here, the pearlite nucleates within the austenite grain boundaries and then the transformation completed with martensite formation.

The experimentally obtained results were close to the predicted values because during carburization at 930 °C for 10 hours, the austenite had enough time to nucleate and grow heterogeneously. Microstructural analysis showed that at the outer layer of the case, almost 3–8% austenite transformed into ferrite at the initial transformation stage and then to pearlite. Almost 40–60% pearlite transformation took place along the grain boundaries of austenite, and the martensite transformed within the austenitic grain boundaries. The image analyzer was extensively used to distinguish between ferrite and martensite phases since application of echant (natal or pickrol) was not helpful. Microhardness tests were conducted to identify the martensite and ferrite since it is difficult to distinguish between the two through visual inspection. The hardness values of martensite (985 HV, 965 HV), pearlite (441 HV, 354 HV) and ferrite (244 HV, 221 HV) were recorded. The decrease in the hardness of martensite indicates a decrease in the carbon concentration from surface to center. The higher pearlite hardness indicates finer pearlite at the outer layer of the specimen due to the faster cooling rate. The lower hardness at the core indicates no martensite formation and is reflected in the results obtained through the image analyzer, microhardness test, and graphical point count method.

12.5 CONCLUSIONS

In the presented work, microstructural evolution in the case of carburized steel was predicted using a simplified mathematical model. The transformation of microstructure depends on the carbon content of steel and cooling rate. Cooling curves were obtained using the FEA (ANSYS) package. The transformation equation of Johnson-Mehl was modified by including supercooling (ΔT). The exponent n does not represent a constant Johnson-Mehl value (range 3–4) but rather varies with time. The theoretically predicted values matched well (between 10% and 15%) with the experimentally obtained microstructural evolution.

REFERENCES

Banerjee, M. K. (2018). Physical metallurgy of tool steels. *Reference Module in Materials Science and Materials Engineering*. https://doi.org/10.1016/b978-0-12-803581-8.09810-6

Bates, C. E. (1988). Predicting properties and minimizing residual stress in quenched steel parts. *Journal of Heat Treating*, 6(1), 27–45. https://doi.org/10.1007/BF02833162

Buczek, A., & Telejko, T. (2013). Investigation of heat transfer coefficient during quenching in various cooling agents. *International Journal of Heat and Fluid Flow*, 44, 358–364. https://doi.org/10.1016/j.ijheatfluidflow.2013.07.004

Hillert, M., & Höglund, L. (2004). Comments on "reply to comments on kinetics model of isothermal pearlite formation in a 0.4C-1.6Mn steel." *Scripta Materialia*, 51(1), 77–78. https://doi.org/10.1016/j.scriptamat.2004.02.043

Hutchings, I., & Shipway, P. (2017). Design and selection of materials for tribological applications. In *Tribology* (2nd ed.). Elsevier Ltd. https://doi.org/10.1016/b978-0-08-100910-9.00008-8

Maher, I., Eltaib, M. E. H., Sarhan, A. A. D., & El-Zahry, R. M. (2014). Investigation of the effect of machining parameters on the surface quality of machined brass (60/40) in CNC end milling—ANFIS modeling. *International Journal of Advanced Manufacturing Technology*, 74(1–4), 531–537. https://doi.org/10.1007/s00170-014-6016-z

Niederschlags, P. (1851). *Ueber Diffusion*. https://onlinelibrary.wiley.com/doi/epdf/10.1002/andp.18551700105

Prediction, P., Quench, B. Y., & Analysis, F. (2009). Quenching of aluminum alloys: property prediction by quench factor analysis. *Quench Al.qxp. June*, 23–28.

Sinha, V. K., Prasad, R. S., Mandal, A., & Maity, J. (2007). A mathematical model to predict microstructure of heat-treated steel. *Journal of Materials Engineering and Performance*, 16(4), 461–469. https://doi.org/10.1007/s11665-007-9041-3

Teixeira, J., Moreno, M., Allain, S. Y. P., Oberbillig, C., Geandier, G., & Bonnet, F. (2021). Intercritical annealing of cold-rolled ferrite-pearlite steel: Microstructure evolutions and phase transformation kinetics. *Acta Materialia*, 212, 116920. https://doi.org/10.1016/j.actamat.2021.116920

Vander Voort, G. F. (1991). Atlas of time-temperature diagrams for irons and steels. *ASM International*. https://books.google.be/books?id=xWGxHU_1XpUCx

Index

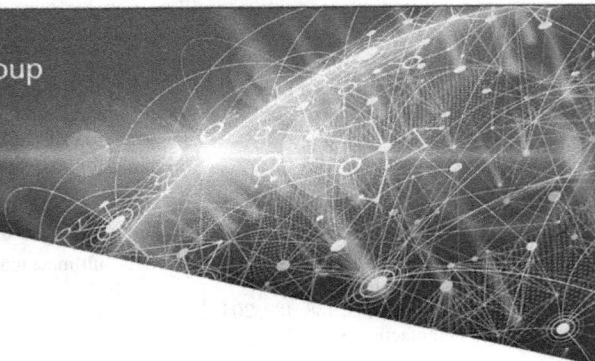

For Product Safety Concerns and Information please contact our EU
representative GPSR@taylorandfrancis.com
Taylor & Francis Verlag GmbH, Kaufingerstraße 24, 80331 München, Germany